Gloria Hillard

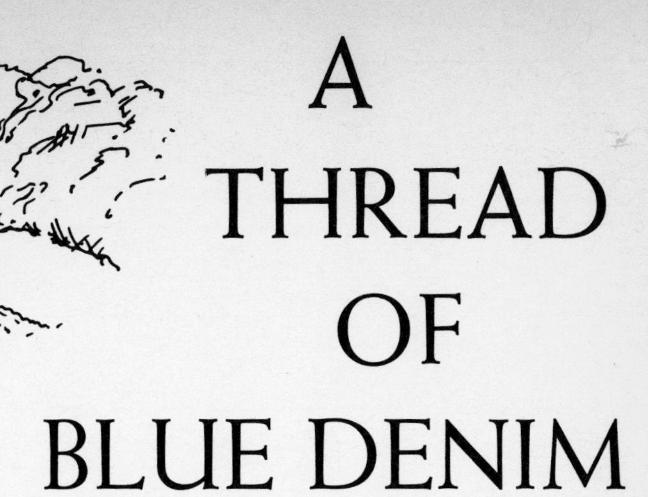

A THREAD OF BLUE DENIM

By Patricia Penton Leimbach

Illustrated by R. Bruce Laughlin

Harper & Row, Publishers, New York
Cambridge, Philadelphia, San Francisco, Washington
London, Mexico City, São Paulo, Singapore, Sydney

The essays in this book originally appeared, in slightly different form, in *The Chronicle-Telegram*, of Elyria, Ohio, and/or *Farm Journal*.

 For information address Harper & Row, Publishers, Inc., 10 East 53rd Street, New York, N.Y. 10022. Published simultaneously in Canada by Fitzhenry & Whiteside Limited, Toronto.

First PERENNIAL LIBRARY edition published 1987.

Library of Congress Cataloging-in-Publication Data

Leimbach, Patricia Penton, 1927–
A thread of blue denim.

"The essays in this book originally appeared, in slightly different form, in the Chronicle-telegram, of Elyria, Ohio, and/or Farm journal"—
Originally published: Englewood Cliffs, N.J. : Prentice-Hall, 1974.
1. Farm life—Ohio—Vermilion. 2. Leimbach, Patricia Penton, 1927–
I. Laughlin, R. Bruce. II. Title.
S521.5.03L46 1987 977.1'23 87-45063
ISBN 0-06-097092-8 (pbk.)

---

87 88 89 90 91 MPC 10 9 8 7 6 5 4 3 2 1

For the Pentons, who indulged me with dreams;
For my sons, who surrender their private lives;
But mostly for Paul, who ties up the loose ends.

THE END o' WAY

FOREWORD

Rugby is a crossroads community that for all practical purposes has ceased to exist. Only the sign at the cemetery still carries the name it bore when it was a political entity.

The red schoolhouse at Rugby Corners is a storage shed now, and the post office house falls to ruin. The churches, the mill, the blacksmith shop, and the general store are gone. Most of the farms have passed from the hands of the old families, but if you take the left fork at the schoolhouse and follow the short road that skirts the rim of the valley, you will arrive presently at End O' Way Farm, and nobody alive can remember when the Leimbachs didn't live there. End O' Way is 240 acres of sandy vegetable land cradled in the elbow of the Vermilion River. Lying between the upland and the river itself are 100 acres of timbered hillside and sloping meadows.

Despite the "Dead End" sign down at the fork, there is life at End O' Way—a life so abundant and joyous I have never been able to keep it a secret. For the past nine years, through a weekly column in the CHRONICLE-TELEGRAM *in Elyria, Ohio, and occasional articles in* FARM JOURNAL, *I have been trying to communicate what it is to be a country wife at End O' Way in the exciting now.*

There was a time when the fabric of country life was mostly blue denim, when a woman was isolated in the country struggling with her washing and mending, mingling only with other blue-denim people. Communica-

tion was only by R.F.D., and her views were limited to those from the clothesline.

But blue denim and farm life have both changed. Blue denim has been embraced by the world at large as a symbol of the down-to-earth values farm life evokes. Farm life transformed through communication is far more involved with the world, and blue denim becomes more a point of view than a way of life.

Deep in the bone marrow of every farm woman there runs a thread of blue denim. It links her securely with a past about which she has few illusions and a future where she will understand better than most the hunger for the real, the honest, and the stable. No matter what her life touches, it will be wound with that thread.

This is our life in Rugby Corners today, where you may discover that the thread of blue denim is often at loose ends. . . .

CONTENTS

A THREAD OF BLUE DENIM

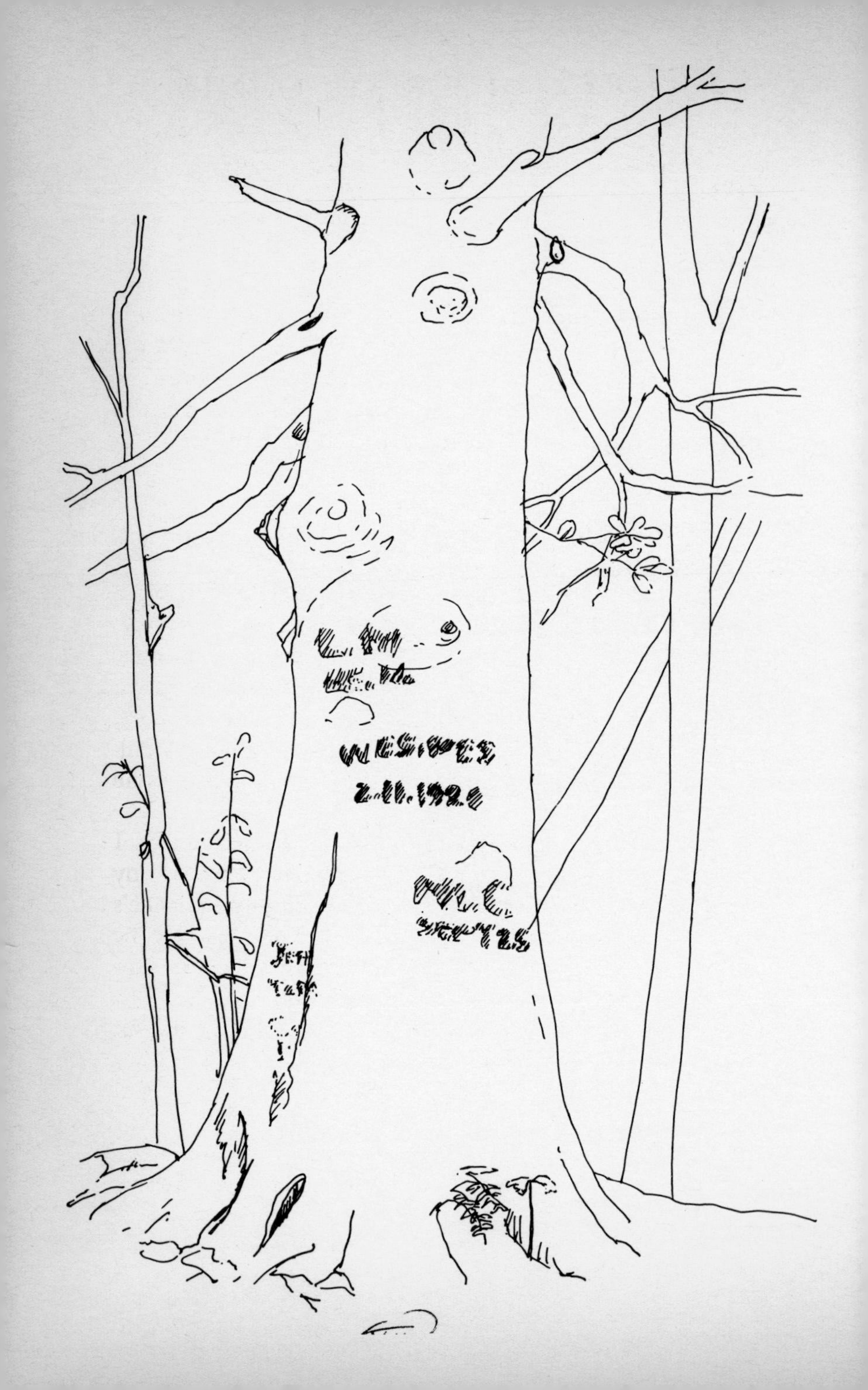

1 THE LAND

The land was ours before we were the land's.

Robert Frost, THE GIFT OUTRIGHT

The land takes possession of us in several ways. It attracts the eye and captures the heart with its beauty; it teaches us the lessons of life in its natural metaphors; it enslaves us with its demands. But ultimately we surrender our souls to the land, because it is abundant in response to our faith.

Fringe Benefits

It would be good to look back and say that we chose to farm because it seemed the best of possible worlds. The truth is that it was the line of least resistance. We were a couple of twenty-seven and twenty-three years, suddenly but securely in love and not too well informed on alternative occupations. Besides, there was a house available with the family partnership, and that seemed some sort of miracle.

These thoughts came to me one August afternoon as I lay on the bank of the irrigation pond and watched my youngest son, Orrin, through a curtain of Queen Anne's lace. All the muscles on his tanned little body flexed as he pulled at the oars of the dinghy skimming across the surface of the pond toward the willow clumps and me. The ripples from the dinghy broke the mirror image of walnut trees; a startled turtle plunked into the water just below my hideaway. What a good life for a boy! A great life! And it was all plain luck.

Up in the wheat field my husband, Paul, was combining, and our second son, Teddy, followed after him clipping straw stubble. I had just come from there, and that scene was part of the satisfaction I felt. There is a studied competence about Teddy's performance with a mower or anything else. He plays no mental games; his mind is there where one sees him, every nerve fiber tuned to the work he knows from the years on his father's lap. As I watched, he slowed for a woodchuck hole, then gradually resumed speed. Further on he threw the mechanism out of gear, climbed down, and carefully unclogged the mower blade. When he pulled up beside me he removed his baseball cap, pushed back his long hair, and mopped his thin face with the back of his hand.

One year Teddy was three, the next year he was forty-three. We offered him childhood and he rejected it. The sandwich and iced tea I had brought were no picnic offering to a small boy, but a ministration to the appetite of a man. And Teddy's manner acknowledged the courtesy, in one of our rare moments of complete understanding. Already at ten he was the kind of son every father dreams of having to follow in his footsteps.

A letter from our oldest son, Dane, was a third source of satisfaction on that day. His summer job in an Austrian factory was like child's play to a farm boy. The workers there considered him too zealous, thought he didn't take enough breaks. His letter expressed gratitude for the breadth of the experience. What surprised me was what followed—a full page of specific questions about the farm and a strong request for an answering letter "from Dad." Details of operation and economics we had thought never concerned him seemed somehow important in his seventeenth summer.

Healthy, responsible, happy, and accomplished boys—so they grow from the freedom and discipline of life chosen for them all too casually.

Americans have always joked about the farm: "It's a great place to come from." We encourage our youth to choose vocations that will bring a sense of fulfillment.

The truth that came to me there in the clover by the pond on that summer afternoon was that the welfare of our children has a great deal more to do with our satisfaction and fulfillment than anyone ever tried to point out to us when we were twenty-three and twenty-seven. The farm is a great place to come from; it is also a great place to go back to!

Hopeless Cause

SPRING IS DEMONSTRATING! HER BARRICADES OF green encircle the house and stretch to the woods beyond. She has hung a red flag of emperor tulips against my white board fence. Forsythia stands tall by the down-spout and stares brazenly into the kitchen to mock my toil at the sink. There is a persistent twitter and chirp, and now and then a shrill insistent whistle of birds. A buzzard above the valley soars in the air currents, swooping, climbing, and soaring again.

Hurrying out to toss my garbage behind the woodpile, I am assaulted by a chorus of orange-throated daffodils. Down against the rail fence amid moss and fungi, an early anemone pleads her private cause. I flee again to the kitchen away from the emotional pull of these demonstrators. The bees, the butterflies, even the wasps are joining in.

The breeze reaches out to me with gentle fingers, but I pull away, back to my rigid disciplines—beds and brooms and laundry. From an upstairs window I see the weeping willow, her long gold tresses swaying in rhythm with the spring call.

Shaking a dust mop, I look off to the field where my husband plows. But no matter the onslaught of spring, I will not yield my agenda to this rebellious mob. A fat black cat prostrates himself in the sun on the doorstoop, but let him! I am adamant.

The children come from school and align themselves

immediately with the revolt. Off with jackets and hats, away with lunch boxes and books. Out with fish poles, balls, and bicycles. Spring! Spring! Spread the word!

And then comes that lone emissary who with quiet enthusiasm and uncomprehending vision wins the battle—a rumpled boy with dirty hands, wearing ragged gym shoes and jeans, wet to his knees. A slingshot made from a forked stick and a strip of old inner tube is hung around his neck, and he slips in to sneak a pocketful of navy beans.

"Hey! Where are you going with those beans?"

"Out behind our clubhouse," he answers. "We're having slingshot practice. You oughta' come. It's nice out."

It's not a very serious invitation, no demand certainly. Just another gentle voice joining those of the anemone, the breeze, the willow, the cat. It was a long day of demonstration and I am weary of resistance. I pull my sweater from the chair, follow him out the back door, and surrender myself to the spring.

Arbor Day

PAUL AND I WERE AMUSED BY A *New Yorker* cover one spring showing an exasperated husband placing a surrealistic sculpture in his courtyard, guided from a judgmental distance by an arty wife waving him a little to the left, a little to the right, and so on. Every married couple looking at the picture must have seen themselves in the scene, he standing helplessly with shovel and tree, she with beating gums and waving arms.

We've gone through it dozens of times. Paul, bent on rapid dispatch of a bothersome task, comes to the back door with a spade and shouts, "Where do you want this thing?" Then we circle the house, size up the existing plantings, and when he mutters, "I haven't got all day," I say, "Right here." I back off, look at it from all angles, and wave in another mistake—for as sure as God made little green trees, we've planted every one in the wrong place!

For five years the pink dogwood struggled on the dry, shady side of the sugar maple, then died. The Japanese maple turned out to be just left of home plate on the ball diamond, and the third spring played its final inning. The mountain ash was directly under an electric line, and if the borers hadn't gotten it, the linemen would have. The Persian lilac is being smothered by a small flowering crab it used to have to holler to, and beyond the crab a basswood is coming up strong and promises to shade them both.

Ten years ago Paul brought a small white pine from the woods and set it at the edge of the lawn, paying no attention to a volunteer maple struggling six feet away. Thirty or forty times through the years I said, "One of those trees has got to go!" Alas, now the maple has foliage on the south and the pine has foliage on the north. They coexist like Siamese twins with a common heart—Paul's.

My mother-in-law was an acquisitive gardener who made the mistake of cramming her vast collection into one small lot. Three magnolias, a hawthorne, a Japanese and a cornelian cherry, a golden rain, a golden chain, and a tamarisk stand tête-à-tête in a high hedge. Each tree could have dominated a city lot.

Our farm manager, Ed Syrowski, has finally succeeded in transplanting a beech. Did he put it forty feet from the nearest tree as he should have? No, he put it next to the Japanese cherry at the end of that "hedge" he took over from my mother-in-law. "I figure we'll all be dead before that tree's big enough to quarrel with the others," he says. But from previous experience I'm inclined to be skeptical.

Who can tell where you'll want a tree in ten or twenty years? It's like a two-year-old choosing a vocation. I was nostalgic about the linden in our backyard at home, and thought I should have one at End O' Way. So we planted a heart-leafed sapling out back several years ago. How could I have realized that when the linden grew as high as the house it would obliterate a view of woods and riverbank that I had grown to love?

It takes a lot of living, indeed, to finally recognize the

potential of a scrawny stick with a ball of dirt clinging to its roots.

F.O.B.—Friend of Beech

A BREAKDOWN IN THE FIELD IN MIDMORNING IS not calculated to improve the temperament of the farmer—it is in fact not calculated at all. Long winter months of careful examination and repair (it says here in the *Farm Journal*) are supposed to enable the farmer to sail right through the spring and summer season without mishap. Very nice in theory, quite otherwise in fact. But to the farmer's wife, those breakdowns often yield golden quarter hours of unexpected leisure.

Happily this morning we are working in one of the fields next to the woods, so while Paul disappears for repairs, Orrin and I escape through the barbed wire and find ourselves in the captivating presence of a dozen huge beech trees that put down their roots well over 100 years ago. Sixty years are charted in the indelible initials of Paul's father and his lifelong neighbor Nelson Newberry. Reason says they should be farther up the trunk than the year the young boys carved them there, but botany and fact say no. At another spot on this same tree another generation can be identified—Paul and his brothers and some neighboring cousins. Now, I note, decay sets in around the roots, and far up, a couple of huge limbs are dead.

"They're dying," I murmur to Orrin as we stretch out on the mossy bank, hands under our heads to luxuriate in the pleasure of being here, pondering these old, old residents. It is their privilege to die, I think, for they have shaded this bank for an age already. Only then will the hundred offspring growing from the vast root system get the chance they need to push up and renew the ageless cycle.

The folds in the smooth gray bark where the branches emerge from the trunk give them the semblance of well-scrubbed elephants; fine lines at each knot and juncture highlight their great hulks. But what is it that makes them so awesome? Is it their quality of openness, of fully revealing themselves in their long, smooth, silvery limbs—unlike the maple and the oak, whose leafy inner growth conceals their secrets? Is it their massiveness? I don't know, but I do know that you can take your troubles to a beech tree and leave them there. And when I meet someone who knows and loves beech trees as I do, I have found a soulmate.

Orrin mistakes my thoughtful silence; he cannot know that I am a mixed-up Druid with a fixation for beech trees. All I have said is, "They're dying."

"Be quiet," he calls to Teddy who just then crawled through the fence to join us. "We're listening to the trees die."

Scentsation

On the infrequent days when a northeast breeze wafts us a strong industrial smell from the distant auto assembly plant, I am impressed that I take my fresh country air far too much for granted. Paul, on the other hand, often surprises me with his sensitivity to scents I never notice.

I gave him an odd look one day recently when he inhaled deeply and with an almost euphoric sigh said, "Smell those potatoes growing!" I will acknowledge that a field of potatoes in bloom has a pleasant fragrance, but plain old green vines? Potatoes are our chief crop at End O' Way. Sixty acres in a lush summer could have a "dollar-green scent," I suggest.

Last night it was the neighbor's cornfield that was transporting him. "Just get a whiff of that corn," he murmured. His favorite smell is the cloying essence of black locust bloom rising from the distant valley on a spring night. Knowing that he planted all 1500 of those locust trees probably adds to the pleasure.

Having been brought up with fruit, I think an apple orchard in blossom is quite the finest country scent there is. And now that I am no longer obliged to lug heavy baskets of tomatoes up and down the straw-mulched rows, the smell released by damp, cool tomato vines seems almost a pleasant thing.

My parents' old clapboard house was surrounded by lilac bushes that stretched to the upstairs windows, and when I want to lose myself in a pleasant memory, I am back in my old bed by the west window sniffing lilac on a damp evening in May. In early summer the linden tree sent its sweet perfume in the south window of that same little room.

There are a number of smells from childhood that are gone like those lilac bushes—the aroma of my grandmother's enclosed porch where the cream and milk separating was done, the smell of apples, potatoes, and eggs mingling with that of woodsmoke, which was everywhere. The smell of Grandma's pantry belongs in that category of aromas that nostalgia renders perfume (someone should capture that essence in a spray can).

But it's a new day and there are new scents that a farm child will remember. On this June morning as I sit on a tractor seat in the bean field, it is not beans or corn or black locust or even clover bloom that drifts in the breeze, but the chemical odor of insecticide. I cannot say that I find it offensive.

The fields are weed-free, the vegetables are lush, the foliage a healthy green. I appreciate how much of those circumstances are related to the smell of insecticide, fungicide, and herbicide. I don't knock it.

There is certainly more to a scent than its initial impact on the olfactory nerve. There's a great deal of the emotional and psychological involved, as any dairy farmer will attest. The smell of "ripe" corn silage, which a "foreigner" can only assess as raunchy, offends the cowman not at all. And a dairyman's wife can block it out altogether if she has a mind to. One of my friends vows that over the odor of silage she can smell the roses that twine about an old wagon wheel in front of the barn. "And I've almost convinced my daughter-in-law that she can smell them too!" she adds.

When you go to a country high school and can look out the windows at pastoral scenes of cows en route to pasture, a cornfield waving in the wind, a meadow or an apple orchard in bloom, it must be expected that there is a certain penance to pay. On the occasion of our local graduation one year, hundreds of "gussied up" relatives and friends poured into the football stadium. It was an evening balmy and beautiful enough to allay the fears of the weather-watching administrators. But blowing in from the southeast was a delicate breeze laden with the essence of freshly spread manure.

Considering the solemnity of the occasion, the community might have been outraged as the expensive fragrances of Chanel No. 5 and English Leather, Old Spice and Canoe were dissipated in the stench. But they took it in their rural stride. Echoing and reechoing throughout the stadium as each newcomer greeted his neighbor was the old cliché, "Ah, get a whiff of that fresh country air!"

Cows in the Petunias

FLOWER GARDENING IS A CHALLENGE NO matter who tackles it, but for a farmer's wife, well . . . Your city friends come out and appraise the situation: "Look at that black soil, and all that rich manure!" they squeal, their green thumbs twitching. They don't seem to realize that in addition to the normal hazards—weeds, insects, disease, drought, kids, dogs—a farmer's wife has problems peculiar to the farm.

When we still had chickens I used to cry to heaven in a blue rage at what they did to my borders with their scratching. And then there were the larger animals. One afternoon we came home from a rare summer outing to find thirty-eight steers wandering about the yard. Starting at one corner of the house, they made the circuit across the patio, behind the border fences, and on around, apparently peering in every window and trammeling with their great feet every flower they came to.

I got out my trowel and puttered at things, trying to restore and replant where I could. In despair I complained to Paul, "Do you see why I get discouraged?"

"Would you like to move to town?" he asked quietly.

It's lovely manure indeed, but let no one fancy that the farmers' wives do not pay as dearly for it as the city dwellers' wives. Often we pay more.

Commandments for Young Gardeners

PUT ON THE ARMOR OF HARD WORK, GIRD THY loins with patience, and thou shalt reap anon the fruits of the flower gardener—sunburn, calluses, poison ivy, lower backache . . . and blossoms.

Seek out some hidden place for thy tools, or verily thy spade shall creep, thy trowel shall walk, and thy hoe shall fly away.

Look to thy friends, for a gardener hath need of many. Cultivate older, wiser gardeners with bountiful cuttings and plenteous advice. Cultivate one friend who groweth no garden, that she may come and say, "Lo, what wonders thou hast wrought in this clay and stone."

Give freely of what thou growest, for that when thy stock diest thou mayest go to thy friends and replenish thy loss.

Prune thy shrubbery as thou disciplinest thy children, early and consistently, or verily it shall grow as unlovely as an undisciplined child.

Delude thyself not that when thou hast a garden of perennials thy work shall be minimal.

Beware of phlox, for yea, they are prolific. Take not

unto thy chimney any trumpet vine, for lo, thy dwelling place shall become enveloped even unto the seventh tenant. Eschew also the kudzu vine, for it is a creeping thing that taketh over.

When goldenrod and thistle, mustard and dandelion grow rank in thy garden, speak not of weeds but of thy great success with field flowers.

When the last weed and the last insect shall have been banished before thy hard work and thy genius, know that then cometh thy neighbors' dogs who trample and thy neighbors' children who gather thy blooms in hot handfuls.

Covet not thy neighbors' gardens; know ye not that even must they contend with thy dogs and thy children? Know ye not that the crabgrass groweth always greener on the other side of the fence?

Take into thy dwelling freely the fruits of thy garden. Delight thy mind sometime with the joy of a single blossom in an exquisite vase. And if thou hast but one perfect flower, consider that thou art well paid for thine efforts.

Look to thy thumb; plunge it often into the rich soil, moisten it with water, nourish it with rich manure and fertilizer, and exercise it with tender loving care. Lo, it becometh green.

"Pick the Dandelions"

WHERE THE FENCE POST LEAVES A TWO-INCH crack at the juncture of the patio and the sidewalk, a dandelion spreads its spiky foliage and pushes toward flower and seed and procreation. If I had wedged the seed into that chink, I couldn't have envisioned how it would glorify the cement and the brick and the white fence post. A few inches away a tame and tailored tulip blooms; who am I to say that the tulip is more lovely for my forethought in planting it there or for her rarer quality?

Coming into my kitchen for lunch one May day, I find a limp heap of dandelion stems and blossoms wilting on the counter, gift obviously of the little cookie eaters who play in my yard. They are still too young to sit cross-legged in the grass and string stems into necklaces, too young to wonder at the "you like butter" reflection of a golden dandelion held under the chin, too young certainly to know that dandelions make poor cut flowers in a kitchen but good salad or heady wine. Yet they are old enough to see a dandelion and name it pretty and pick it for a friend. Best of all, nobody screams, "Don't pick the dandelions!"

Crossing the yellow-green of our spring lawn, Orrin bobs down and up, down and up to snap a cap from a dandelion and toss it to the wind. Nobody admonishes him for picking the flowers.

In the two acres of unused pasture that slopes to the pond behind the barn, yesterday's dandelion blossoms are tomorrow's seed spheres, delicate and perfect as snowflakes above the clover and quack grass. The children pick them cautiously, hold them to their lips, blow gently, and watch the tiny parachutes float free of their fragile mooring.

At the end of the pea patch where I pass it on the tractor, another single blossom stands out against the sand like a lonely exclamation point. Beautiful dandelion, flower of childhood! Beautiful in the meadows and the lawns and along the roadside; beautiful in the sight of children and of God.

My little neighbor asks me, "Do you have to plant dandelions, Aunt Pat, or do they just grow?" My heart trembles in the realization that on my answer hangs one of those value judgments by which adults destroy the free acceptance and appreciation of many things beautiful. I could have given her a lesson about weeds, but I have learned that some of the rankest weeds are planted by adults in childish minds to stifle their native inclination to the truth.

"No, Lynn, you don't have to plant dandelions. Aren't we lucky they just grow."

A "Glad" Friend

I HAVE NEVER LIKED GLADIOLI. THEY HAVE A stateliness that I have found forbidding, a restraint in their blooming that my unrestraint defied. Worst of all, they are too often found in bunches, fuchsias clashing with reds and oranges, like a group of beautiful women whose profusion has the effect of making beauty tiresome.

But at last I have made friends with one, a solitary glad in a crystal vase here on the kitchen table. As the Little Prince tamed his rose in St. Exupéry's profound book, I have "tamed" this glad. I spaded the earth and planted the bulb, drove away the hostile weeds, and carried water that she might flourish. I scoffed at myself knowing that this labor would not afford the joy that such diligence deserved. Then came the unexpected thrill of seeing her blooming among the green spires.

Now here in my kitchen I admire her, and her delicacy and restraint at last strike a responsive chord in me. So much of life is here on this stalk—birth and maturation, fruition and death. And thus this friend reveals herself gradually, like every other friend I've known. It is sad to think of the joy I should have missed had I never come to know this one gladiolus well.

Sad, to think that people, like glads, are all too often judged in "bunches" rather than on their individual merit as fruitful, blooming stalks.

Creakings From an Old Crock (while hoeing)

ON TERRITORY TENANTED FOR CENTURIES BY Indians, it is not strange that Indian artifacts occasionally come to light, to be carried in by eager collectors. Paul prizes his collection of arrowheads, skinning knives, axes, etc. For years I have watched the soil as I walked, picked, hoed, or rode a tractor, and have never been favored with so much as a flint chip.

All I ever find are pieces of broken dishes. So not to be outdone, I have started a shard collection. Shards, archeologically speaking, are pieces of pottery by which a civilization can be read, basic indications of how and when a people lived.

My collection, while interesting to me, is of questionable stature. To a science dealing in millennia, it is scarcely embryonic (an archaeologist who dug here gave me a shard that he says carbon tested at 300 B.C.). Yet I like to cogitate, as I hoe, on the story that might be told by my odd pieces of crockery (an archaeologist, I have observed, speaks with solemn conviction about matters that seem to me conjecture at best):

. . . Judging from the fact that these shards were found so far from the remains of the dwelling, we may conclude that these people were of an agrarian culture. It is probable that many of these pieces passed through the digestive systems of animals, having been fed them in the table refuse then scattered in the fields with the dung. (We know therefore, that farmers of this era already knew the value of organic fertilizer.)

The printing (called a hallmark) on this piece of china identifies it as having come from an ancient land called England. It is probable that its owner was of the upper classes or had wealthy ancestors who imported their china. Some unfortunate servant girl may well have been dismissed for letting slip the plate that produced this shard.

This piece of brownish-orangish pottery is a rare find. Though dating much later than the watered blue and white with the English lettering, it marks the beginning of a significant era in American history—the "giveaway." These pottery pieces came buried in oatmeal boxes and were treasured by parents as an inducement for the young to eat the sticky gruel. Hence we know that these people survived on a diet of oatmeal, though some of them found it unpalatable.

These numerous pieces of similar scenic design also belong to the giveaway period and are classified as "early A&P."

We reason that this bit of worn glass was broken from a jug carried to the workers in the field. Similar shards are often imprinted with a legend that has given our etymologists some difficulty: "90 proof." We have been unable to discern exactly what it was proof of.

The heavy pottery of plain white is one of our best indications of the personal habits of these people. It is broken from a vessel known as a slop jar, a sanitation facility predating the flush toilet. The word "Ironstone" is imprinted on many pieces of this type. We understand this to mean that the vessel was heavy as iron and cold as stone. . . .

Oh, don't ever waste your sympathy on a lady with a hoe. Her mind is much busier than her hands!

Unhappily Ever After

One evening at a college alumna function I met a long-ago friend. When I told her I was a farmer's wife, her eyes shone and she told me her wistful story:

"Yes," she said, "I bought a farm. Neither of us had any heritage or experience of a farm, but something hit me and I had to do it. My mother made earnest apologies to my husband, 'I don't know what it is . . . she was never like this.' We sold our town house, took our six kids, our dogs, and our lifetime of possessions to the small farmhouse on 140 acres of magnificent valley. Like the proverbial valleys of time and history, this valley lapses in an earlier decade.

"With poor TV reception and a nine-party phone line it is possible in this era to be virtually isolated on a township road in a valley. And yet we were part of something we'd never experienced—community. If we failed to be in church there was a phone call from one of the twenty families asking if we had trouble. People we'd known only weeks had genuine concern.

"For seven grueling and wonderful months we stayed. Drew drove back and forth to his sales job and struggled in between with the nine-party line (yes, we could have a private line for $1400 and advance monthly payment). The children and I learned to farm. Oh, my, what we learned!

"The neighbor who sold us the cow gave me milking lessons. We got a horse about to foal and midwifed the colt. I raised forty pigs. How I loved those pigs! Pig prices were very good when we got in, very bad when we got out. (Drew says it's me; I have the same luck with the stock market.)

"And the work! I didn't know what it *was* to work. 'Make hay while the sun shines' was just a phrase, but now . . . what an overwhelming job it is to work on a wagon behind a baler.

"We never realized somehow that you must buy your seed and fertilizer, plant and harvest your crop, feed it into something, and sell the 'something' before you have any return on your investment. The Internal Revenue will never believe our income tax.

"But the children, how they grew with it. From the moment we went there they were different, especially the boys. I suppose there's a place for a girl from the time she's born, but a boy needs something. In the city you make jobs for boys and set up rules. But it wasn't that way on the farm. They had a 'place and a job that was meaningful. They could see that neglecting their job caused something or someone to suffer. Rules became irrelevant. Working with their animals, seeing them born, nursing the weak ones, they felt themselves part of the life-and-death process.

"At the end of one long, hot day in late summer we ran

out of water (precious, precious water). I sent John down the road to the neighbor's with the tractor and a milk can. I'll never forget the wonderful look on that boy's face when he came back. He had done something vital for all of us, probably for the first time in his life.

"And always there was the view—out of every window the woods and the fields, and to know that we were part of it all. It's indescribable.

"But it was no use. I thought I wouldn't mind—Drew's endless commuting, one bathroom (after having four), almost no water, most of our possessions stowed in the barn—but I did. So we're back in the city, having brought our belongings covered with hay chaff, fertilizer, and bird droppings, and our children, dispirited, back to the rules and the meaningless activity. But nothing's the same. Now we know what the void is.

"We go down on weekends. We're displaced persons, perhaps; we're not there, and we don't seem to belong here. With all his economic, business, and sales acumen, Drew's consuming ambition is to raise 125 bushels of corn to the acre. Yes, I bought the farm and it sold itself to Drew.

"But it's hard living here, wanting and needing to be there. I don't know, there must be some compromise. . . ."

My heart aches for them. There's much I could have told them before, but now it's too late.

The land is a passionate lover; she commands your time, your money, your devotion. She rewards you always with beauty and delight, but she makes no promise beyond that. If you are faithful she will feed and clothe you, but the land is a jealous mistress and will not be trifled with.

Summer Sabbath

THERE IS A PORTION OF A FARMER'S HARVEST nowhere recorded in ledgers or tax forms. Yet it is the portion most inclined to keep him on the farm fighting insects and fungi, the highway department, the zoning board, and the acreage quotas. It is the harvest of satisfaction gleaned in the quiet twilight of a summer Sunday, when like absentee landlords he and his wife go hand in hand to the fields to see what God and the day laborers have wrought.

The beans could use a little water to hurry them along. The strawberries need hoeing. A few aphids feed on a vine at the end of a potato row. The early potatoes must be hilled this week. The wheat begins to yellow and the stalks arc as the grain grows heavy. Thought must be given to oiling the combine soon. All this is swiftly noted and dismissed. These are the cares of the pair who return in the morning, not the concern of the Sunday evening overseers.

What they dwell on is the beauty and the peace and the fullness of their "fortune" on this farm. More satisfying to the senses this evening is the waving of the wheat, the undulation of this green-gold sea. Was ever anything written of a wheat field that does justice to its beauty?

The potatoes—how luxuriant they've grown with the cool damp days. How lovely the white blossoming vines stretching in their straight rows almost to the sunset. In the hedgerow along the creek the elderberries bloom, emitting a musky fragrance. The melons are sending out runners, and here and there a tiny green globe swells behind a withered blossom. A late field of green beans just now breaks the surface of a crusted plot, humbling in the reminder that nothing man does here equals the miracle of the seed.

The breeze dies and there is only a whisper of light in the west as we turn our faces to the night and come down the lane to the darkened house. A feeling of satisfaction and joy fills our hearts and breathes a benediction on the week.

Afterglow

If you are very lucky, someone will interrupt you on an October afternoon in the midst of the raging, roaring, rolling torrent that is harvest, saying, "Let's go to the woods." You'll think of the 300 things you were going to do in what's left of this day and say, "Splendid idea!"

A cover crop of rye blesses a vegetable farm with spectacular green in early October. A line in a favorite poem —"Green, green is a God color"—seems to fit it. Across this God-colored field we go, my companions, my dogs, and I, then onto a meadow gone to weeds. It's a smoky haze of a weed that a farmer's wife can only appreciate in a mood of abandon. I reach down and spread my hand over the whisper of it.

Woods companions must be chosen with care. It is expected that when you pause and look up in wonder at long light filtering through feathery pine needles they will not stumble over you, but join you in awe; and that no one will spoil the miracle by speaking of it. They should know intuitively that you revere this section of woods and share it with them because they're special.

We share a leaf of sassafras pulled off and crumpled and held to the nose for its discovery of lemon. The smell invokes the chill of March, the digging of root, and the brewing of pink tea in warm farm kitchens. It plants the temptation to return in two weeks when this will be a burning bush.

Stepping over small obstacles, we make our way through underbrush to the river bluff, to a point that falls away gradually to the water. Far below, the river is a placid creek with no more motion than if it were a canvas by Rousseau. A yellow leaf that travels its surface is the only indication that it moves at all.

The trees on the opposite side of the valley take on the

hues of autumn. The fellow who holds deed to those woods thinks of them as his own, but today at their loveliest they belong to us. Here on the sloping surface where we stand is the memory gift I will give these friends to carry from this afternoon, an oak sanctuary carpeted in great masses of moss, newly green with the reviving rains of fall.

Here one must kneel and spread one's hands lovingly on the soft green. One friend inclines her head to inspect the star-shaped moss plants, and there is more of prayer in the posture than in all the liturgies of formal religion. Picking up a sprouted acorn, I have a quick vision of the great tree whose miraculous beginning lies here in my palm, a sad awareness that I will not live long enough to see this tree grown.

One cannot absorb enough of this place, and must go away still longing after loveliness. The descent to the valley, the smooth fallen logs, the shamrocks growing in shale, the hemlock beneath a waterfall, the shadows on the still water, nothing satisfies that longing. . . .

It is into that vacuum in the hours and days ahead that the joy of this day will slowly seep.

Twilight for a Dog

PEPPY IS OLD AND THE NAME THE CHILDREN gave her as a pup is ex post facto. Once she leapt on everyone who came into the yard, paws on shoulders, tongue licking. It was her way of demonstrating enthusiasm, but it terrified the timid and the small, outraged the fastidious. Now she is content to lie on an old rug in the sun and merely lift her head to humankind to suggest a welcome.

In her twelve years she has been all the things a good farm dog should be. She was the gentlest of playmates; when the boys were babies they napped against her warm flanks, fondling in their fingers a black velvet ear. As a watchdog she succeeded by ferocious fakery. Guests often had to be coaxed to leave their automobiles. Though we kidded ourselves that she could tear an attacker limb from limb, we knew that so far as trusting people was concerned, she was as naïve as we were. Yet many were the times when her insistent barking alerted us to situations needing attention—roaming steers, a loose pony, uninvited hunters.

Peppy has a son Buddy, who has been her persistent companion through the years. No one watching the two swimming ponds, racing across meadows, wrestling, barking, then finally collapsing exhausted together could but envy the joyous freedom of a "dog's life."

Peppy has been an indefatigable companion to all of us trotting hour after hour behind a wagon or a tractor, keeping pace in the bean field or the melon patch, pausing for moments at a time in the shade of the vines, maintaining ever the image of man's best friend.

But it is as a hunter that she has best justified her existence as a farm dog. Never housed and pampered, seldom fed, she and her son have survived by their native instincts, ferreting rats and mice from the barns, stalking the fence rows for rabbits and woodchucks, supplementing their diet with steer feed and ear corn. Their bodies grew tough and their coats sleek.

There's a quiet pathos about an old dog, especially a spayed female. Forbidden an old woman's luxury of complaint, she merits the respect accorded the true martyr. Her muzzle is white, her muscle gone to fat, her movements heavy and slow. The small pup from next door frisks about teasing to play, and Peppy seems to beg deliverance from the intrusion.

On a rare warm day she still seeks us in the potato field, hungry for the old companionship. She ambles painfully across the rows, her rear flanks at war with the front. The nose is as keen as ever, and suddenly her body takes on the old "point," the rigidity that precedes the chase. But no more the fleet dash, the yelping pursuit, the savage capture. The young hunter in her holds the nose and ears erect, but an old woman whispers, "You are old, dearie." And slowly her rear quarters sink to the ground. Gradually the head also surrenders its hunter's poise, but she continues to gaze in the direction of the scent. I don't know what she thinks, if she thinks. But I could weep for her, remembering the joyous way she once bounded in quest of prey.

A long while she watches, then limps back to the house. Later I notice that she has returned with her old ally. Peppy sits patiently beside the cornfield and her son Buddy (two years her junior) hunts within. There is a yelping of conquest and Peppy proceeds painfully to the feast.

There is great solace, I would remind myself, in having a son.

Head Goose

A FARMER DEVELOPS A CERTAIN NONCHALANCE toward natural wonders. But Paul has never become inured to the miracle of a flock of geese in migration. He can always afford the time to stop and watch.

The mist still hangs in the valley this morning as twenty-five or more birds come honking over in an undulating V, and he halts the potato-digging operation to call our attention to them. How many centuries of accumulated human knowledge went into the development of a radar system such as these awkward birds employ instinctively?

As we watch them pass over, two or three maverick followers on either side align themselves behind lesser leaders to form small V's off the main formation. Out of the northwest they come, and their steady flapping carries them beyond the horizon to the south.

It happens to be Election Day and my mind is very full of the subject of leaders, so I ponder this flock long after it has disappeared. How, I wonder, do they choose the head goose? Is he born head goose, or does he train for the position? Is it every gosling's dream to grow up to be head goose? Is he the most beautiful of the ganders; does he have the lustiest honk? Is he the gander who charms the most geese or nudges the most goslings under their wings? Is he a he at all?

How would I choose a head goose if I were a goose and obliged to choose? I should certainly want to follow the wisest goose, the one with a sense of goose history who knew when to fly and where, the one who could identify predators and cope with them.

I should want a compassionate goose for a leader who was considerate of the geese with the crippled wings but never lost sight of what was best for the good of the flock. I should want to follow a goose who did not drive us beyond our endurance, who guided us down safely to sheltered lakes where we might renew our strength. I would want a leader who saw that we didn't eat so richly and so long that we could no longer fly.

As all these thoughts pass through my head, another formation of geese, a much smaller group, appears from the south, the direction in which the others have gone. Perhaps they are a splinter group at odds with the normal southerly flight plan of the larger flock. They fly off to the northeast honking loudly.

The most important question to consider in choosing a head goose, I suddenly realize, is: Is he headed in the right direction?

The Reapers

THROUGHOUT THE COLD, GOLD DAYS OF A diminishing autumn the farmer pursues the harvest. At Miller's and Dodd's it's apples; at Aufdendkamp's, cabbage and cauliflower; Don Northeim pushes on through acres of soybeans; the Schmalz brothers move into their cornfields, while the Leimbachs work their way up and down the potato rows with their green machine.

At 8:15 Jenny Shinsky comes, quietly opening the back

door and slipping an apple strudel onto the kitchen counter. Ah, that strudel, it will make lunch seem another hour distant. And Jenny! What a worker! She's as much a part of potato harvest as the geese that fly and the leaves that fall. It's more than a job with her; it's a cause, and she suffers along with Paul in the fickle circumstances of weather and digging conditions.

Happy is the farmer who can muster the forces of his neighborhood, coax the women from their kitchens for a few intense weeks of labor. And this is Paul's fortune. By 8:30 they have sent their children off to school and set some meat to thaw for supper; looking very unisexy in hoods and jeans and stocking caps, they are ready to spend the day sorting stones and clods from potatoes. Paul adds relish to his day by flirting with them.

My job is to drive a tractor and wagon alongside the harvester to receive the potatoes elevated out on a moving web. Being married to a man for twenty years doesn't guarantee that you can drive in tandem with him all day (on tractors with differing gear ratios) without a few skirmishes, mostly verbal. But it does enable you to take his loud complaints without bursting into tears.

The rows are long and the driving often cold and tedious. I spend a lot of mental time vagabonding in southern France. I think back through great books, relive the joyous hours of my life. I summon the memory verses that my good English teachers foisted upon me in less enthusiastic days. Or safely outshouted by the tractors and the screeching machine, I sing my collection of song lyrics. I study cloud patterns and the flight of birds. I look for sermons and signs in the weeds that grow and the seeds that blow, and I write great epics in my mind.

If there is a lull between wagons or a time-consuming breakdown, I climb up on the machine and chat with my neighbors. When we have exhausted conversation, I read a few paragraphs of the pocketbook I carry along, or flop down on the friendly earth and catnap.

The summer vegetable ground rests beneath a blanket of rye. Later plantings are still distinct rows that stretch off toward the valley, foreshadowing the spring. The mailman drives up the road, and then comes the kindergarten pool. The cluster of houses and barns in the distance is a warm oasis of lunch.

The gleaners come to salvage our loss and fatten their larders. Their unconditioned bodies struggle with grocery bags and gunny sacks and washbaskets full of potatoes. The bending and lifting and carrying are alien exercise, and they sink to rest on their sacks or crates. For all their world of chrome and plastic and polyester, I see them as the sons and daughters of Millet's French peasants. And when they stop our apparatus midrow to pay for their estimated lot, there is a satisfaction in their faces that is rarely seen these days.

The southern ends of the rows lead to the river valley, and as we make our turns we get warming glimpses of sunlight through sycamore. But the continuing miracle of the day and the season is the potatoes, load after precious load of rich, damp yield. So sharp yet is the memory of fresh spring mornings, sitting up behind the planter watching angular seed pieces drop sensually into the opening furrow.

We were there in the beginning, Paul and I; here we are at the end. It would be easy to forget how much we owe to technology in between. The irrigation pipe stretches in silver ribbons to the east to remind us that this could have been a different story, as it often has been.

Late in the afternoon come the school buses up the road, and thoughts move toward home and supper.

The family farm—less and less of the nation's agriculture depends upon it. It's neither as cheap nor as efficient as an industrial farm complex, but it is very human and much more satisfying. Much as the agribusiness complexes, it utilizes science, technology, and people to produce food. But it does not destroy the individual or subjugate the mind, as the husband of this wool-gathering tractor driver will freely testify.

Finale

PLODDING THROUGH SOME THEOLOGICAL WRITING recently, I found myself stumbling over the word "eschatology," and was somehow pleased with what the dictionary had to say: "Dealing with last or final thing. . . ." Too good a meaning to confine to theology, I thought at the time.

Today as a raw wind blows a freezing rain and the weatherman makes his dismal prediction of "temperatures in the 20's tonight," I think of that word again. Surely this is an "eschatological" day. . . .

No more shielding the geraniums with a fold of newspaper; today they must come out and be hung in bunches by the roots. The tomatoes must be picked, red and green alike, and stored away from the heat to ripen slowly. Today the sweet potatoes must be snipped off at the ground so the frost will not penetrate the roots. And the last of the flowers must be gathered and brought in big armfuls to the kitchen where they will be wondered over tomorrow (how many flowers can one house accommodate tastefully?). There is only time to put the glass panels in the storm doors and to regret last week's neglect of the storm windows.

The men have more to worry about than storm windows, as they collect the steaming vehicles and pour in antifreeze. They make excursions to the far reaches of the farm, draining irrigation lines and checking pumps, remembering with a shudder pipes previously overlooked.

There is a frantic concern to "get things in out of the weather"; a search is made for electric heaters and extension cords. Kerosene and fuel oil cans are carried around to places where they may be needed. In the midst of it comes a phone call that the church roof is leaking. It would seem that the Lord could take care of *something* Himself today!

On such a day a man wonders how his friends and neighbors are doing, if their perishable crops are in, and finds time to offer only his prayers.

There are chilling memories of scratching futilely in cold wet fields after potatoes that past winters subsequently claimed.

Crossing the yard at twilight, it seems like a good idea to gather the sticks blowing there, to retrieve broken baskets from the hedges and carry them to the furnace room. It is a time to rejoice in building a furnace fire and taking an odd pride in providing one's own comfort. Now, at dusk on this "eschatological" day, comes the hour to brew a cup of tea and make it a toast of thanksgiving.

Finally on Thanksgiving

"FIRST THERE WAS THE SHIP, AND THEN there was the storm. And the first baby was born and they got off on the rock. And then they met the Indians and got the corn, and then they had the first Thanksgiving," wrote Teddy when he was six. Thanksgiving at six is pictures of people strangely dressed in black and gray with buckles on their shoes.

It is the day the relatives come with their damp, unwelcome kisses, the table is stretched to fill the dining room, and dinner is forever getting served. I remember it all, and the changing impressions . . . wishing I were old enough to sit at the big table instead of the smaller one at the side, wishing that my pine-cone turkeys would stand more firmly on their pipe-cleaner legs.

Setting the table carefully with the heirloom linens and the stemmed goblets was a joy of growing older. The Haviland plates were taken to the kitchen to warm, and the hand-painted serving pieces were set out to receive the food. It mattered little that the china didn't all match, or that the goblets were chipped. In some odd way it assured me that our "peasant" existence was only a facade, that behind it all we were gentry born to the silver spoon.

When the food finally came and we assembled about our round table, there was a smug sense of unity and well-being. The plates were never big enough, and the gravy ran over the corn while the pickled peach rolled into the mashed potatoes. Everyone exclaimed about so much food, and where would they put it all, and why couldn't their stomachs hold more.

Thanksgiving in the teen years was washing dishes for hours, then trudging through wet pastures with a swain who thought Thanksgiving was a hunter's holiday. Eventually Thanksgiving was homecoming and bringing your roommate. It was looking for the action in the home town, and working on term papers, and going back to college disappointed but never admitting it.

Having a fiancé at Thanksgiving was finding that suddenly the house was too small and the family too big and too noisy, and there was no place to be alone. Then Thanksgiving was "over the river and . . ." down the freeway to Grandmother's house and taking your place at the stove and understanding at last what took so long.

And all through those years, giving thanks was something superimposed on the holiday at the last minute when the turkey steamed on the platter, like a salt shaker you've forgotten and jump up to bring to the table.

If, however, you are among the blessed who cling to the land, there comes finally a year when the full realization of Thanksgiving is upon you, when you see back through the harvest of autumn to the cultivation of summer, to the tilling of spring. You hold the fruit, sense the plant, and bless the seed.

You kneel before the altar of weather—rain and sun and wind and frost. You rejoice in technology; you sing hymns to good health; you praise God for a husband, children, workmen who are skilled and faithful in their labor. You thank the Lord for food and the privilege of producing it. Now, at last, you understand about the ship and the corn and the people with buckles on their shoes. Finally you have a thanks-giving—and on it you superimpose a holiday.

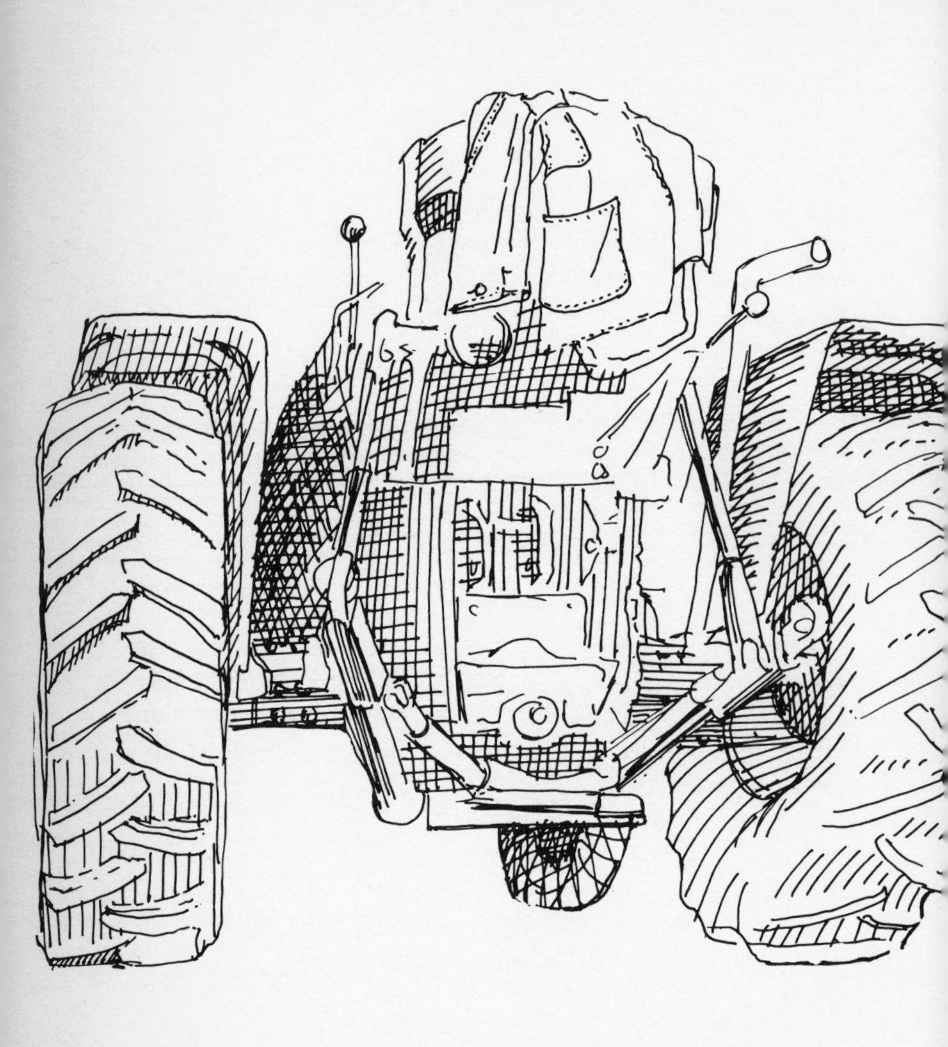

2 THE FARMER

. . . A bold peasantry, their country's pride . . .
Oliver Goldsmith, THE DESERTED VILLAGE

There is an inherent integrity in a farmer. To live with one and give yourself openly to all that involves is a moving adventure in growth. To appreciate his impatience with triviality and his nonchalance before obstacles is to understand his reticence before the world. To watch him through the years in his preoccupation with weather, his quiet appreciation of the natural order of things, his relentless devotion to his land is to discover, wonderfully, the primordial essence of man.

Winterlude

Ask a fellow on the first of July why he became a farmer and he's likely to tell you it was simply a mistake he's never overcome. Ask him the same question at the end of February and he'll come closer to the truth: "It's a good life."

He has struggled with depreciation and personal property, with income and outgo and sent off his tax check. The bills and receipts are filed in a shoe box in the attic. He has assessed the store of corn, wheat, and hay still in the crib, the bin, the barn, and is amazed to find himself solvent. Then he slowly but deliberately proceeds to alter that situation.

He makes "luxurious" trips to the elevator, the hardware store, the tractor dealer's to complete the deals that have been pending for the needs he anticipates—an auger, a plow, a planter, perhaps a combine. The salesman who appears at the kitchen door is patronized as never in August. He's sent back to the car for brochures, cross-examined, then told in fat-cat tone, "We'll let you know."

When the farmer's wife says, "How about tiling the bathroom?" he doesn't change the subject as usual. He may even glance through the paper and casually remark, "They're having a sale on floor coverings."

Coming in for tea in midafternoon, he gazes around at the walls and woodwork as though he hasn't seen them in quite a while. Then they talk, he and his frau, of improvements that could be made. Or one day he takes a monkey wrench to a leaking faucet, and before he knows what's hit him he's remodeling the kitchen.

There are other satisfactions that wouldn't occur to him in midsummer. He's not doing anything today that he couldn't put off until tomorrow. He rather enjoys answering the phone at 10:00 A.M. and hearing a startled friend say, "What are *you* doing in the house?" Following an afternoon nap he cleans off his cluttered desk, peruses catalogues and magazines for new varieties and ideas.

When he thinks with a sigh of the nine long months framing this interlude of peace, the balmiest day of spring comes to mind. The scent of damp earth is in the air, far-off sounds seem somehow near, blackbirds dart about building nests in the mustard; and the fellow who winds back and forth with tractor and plow seems more lordly than bonded. It's a prospect to shut out the heat, the sweat, the frustration of so many days to follow.

But for now—he sits in his warm kitchen surrounded by diagrams of his fields; he lays out his crops, estimates his seed and fertilizer, muses over the innovations he'll adopt. Anything seems possible—this is a very good life indeed!

Love Affair

When you wake with a start on a still March morning to find the bed beside you cold and empty, it's good evidence that the lord of the manor has renewed his "affair." It's the first of many rendezvous that will keep him from your bed late and early in the months ahead.

He stops by the kitchen at 7:30, his eyes shining with love for his mistress—the land. No matter with what relief of body and spirit he cast her aside in the fall, he finds her fresh and young and capricious on this morning in March, and he is a boy again joyous in love. The primordial something that stirs the birds, that takes the swallows to Capistrano and the wild geese to Canada, has spoken to this farmer in the night. "Be up, young lover, and sow your clover among the wheat while the field is firm and the breeze sleeps." And he answers the call. . . .

He shivers now in his hood and coveralls, and you brew him something hot while he shares the secrets of his land discovered during this first loving spring encounter. And then he's off to finish before the breeze springs up to spoil the task.

The freezing and thawing of early March have honeycombed the soil, leaving crevasses down which the rains will wash the clover seed to germinate among the roots of the wheat. When the wheat and straw are harvested in late summer, lo, there will stand the meadow for next year's hay crop.

The children used to weep when they would come down to breakfast and find that their father had been out doing tractor work without them. And I would explain with unusual firmness that Daddy always plants clover seed very early in the morning and alone. They will learn, as I have learned, that when a man keeps a lover's tryst, he keeps it alone.

Oil!

When a "slicker" drives up the lane in a new automobile with Western plates and climbs out wearing a ten-gallon hat and leather boots, a farmer knows he's about to be approached about oil leases. He might as well turn off the tractor. There's going to be a lot of gassing —and he better save his.

This Westerner will start the conversation with an intelligent observation on your farm operation (his "daddy" was a rancher). And the farmer will give him the benefit of a few minutes of up-to-date farm know-how. (Every farmer sympathizes with that spirit in his fellow man that never completely relinquishes its ancestral claim to the soil.)

Finally they'll talk of oil. The landowner, foolishly he knows, feels the surge of excitement and importance that belongs to the saga of oil. Through his mind flash vignettes of himself as an oil tycoon. His eye falls on his wife seated up behind the potato planter looking like an enormous teddy bear in her insulated underwear, and he features what Neiman Marcus could do for her.

. . . They're a stylish pair with stylish children sailing before the wind in Montego Bay, or skiing at St. Moritz. They build the house they've always dreamed of over on the riverbank. They trade in the Ford for a big Mercedes, and send the boys to Princeton and Dartmouth instead of the state university . . . castle on castle.

". . . A dollar an acre—$260 a year for the whole place."

He is back in a flash from St. Moritz. The castles vanish, and reality takes possession. Two hundred and sixty dollars they'll give him for his right to privacy, for the potential invasion of his cherished land with who-knows-what manner of destruction!

He glances around him at the green rye and the brown earth, at the deep woods up and down the bank, at the broad pasture on the hillside where the cattle graze. He looks down to the white frame house where his grandfather came as a boy, where his father and he and his sons were born. He looks at the three big maple trees that shade it, trees that represent for him three brothers, three sons.

"Nope," he says, "don't think I'd be interested."

"This is the Way the Gentlemen Ride . . ."

ALONG WITH THE INCREASED COST OF FARM machinery these days there comes a little compensation in thrill value. The new breed of tractor, for example, is a great bellowing behemoth as high as a room and as powerful as Paul Bunyan's ox, Babe. We have one. It cost more than Great-grandfather Leimbach paid for the whole farm eighty-five years ago, and we may never own it free and clear. But you sit up in that padded armchair, survey those fancy dials, press the button that kindles all that power; then you put it in one of its sixteen speeds, release the clutch, give a flick of the finger to the power steering, and feel the whole earth tremble beneath you. Who cares what it costs! You are a lord—nay, a king. No lord ever sat in such comfort wielding such power.

It is probable that if you let the banker come out and merely sit on it once a month he would cancel the interest payments. Let him take a spin on it, all 300 horses galloping, and he'll trade you his bank, his split-level, his country club membership, maybe even his wife.

I say we have one. I don't want to convey the false notion that I drive it. I just climb up there every once in a while to get the feel of where all the money goes. The minute I put in the clutch and push the starter button, people converge on me from all directions, impressing on me that I'm incompetent or, worse yet, that driving a machine so large will be detrimental to my feminine image.

Image, ha! We do have "his" and "her" tractors around here, but you can bet your monthly milk check that "hers" doesn't have a padded seat with arms and backrest and power steering. It's more likely to be a hand-crank job with a damp burlap sack to sit on.

"Hey, Dad," I said one day, "why don't you get me a seat like that for the back of the potato planter?"

"For seventy-five dollars! Are you out of your mind?" Obviously, where luxury is concerned it's a question of whose heap of flesh is on the throne.

But I don't begrudge Milord his padded seat, his 300 horses, his power steering, his elevated perch, or his computer dials; for clearly he's a new man with a whole new self-image. Fitting ground up along the back road, he watches the sports ride past to their golf matches; he pushes the throttle forward a little, settles more comfortably into the foam rubber, sets his nose very high, and roars down the row.

If he acts a little officious when he comes in for lunch, it's understandable. When you are lord of the earth, creamed tunafish on toast must seem like shabby peasant fare. But Milady coddles him and sighs romantically as she sends her swashbuckler back to his "chargers." Then she carries her account books to a window where her gaze follows him proudly during the long afternoon of balancing books to meet mortgage payments.

Sweat Shop

Paul is proud of his potatoes and his green beans, his corn and his melons. But the finest products he produces year in and year out are good workers for the labor market, dozens of them every year. In May my phone will begin to ring and small voices will ask if they can have a job this summer, kids eleven, twelve, thirteen, or fourteen. If they tell me they're sixteen, I say, "Sorry, you're too old for this job." By the time a kid is sixteen it's too late to make a bean picker of him. He has already learned that he can make more money for less work at almost anything. But if he starts at eleven and earns $1.50 a morning, he has really accomplished something. If he sticks at it for three or four years, you can bet your bottom dollar that he'll never be on the welfare rolls. As I tell my little bean pickers year after endless year, when you have submitted to the discipline of bean picking for a summer or two, every other job you take on will seem easy. "You have nowhere to go from here but up!"

At twelve the comradeship and mischief keep you going when the rows seem endless. You discover one challenging morning that you really don't need to lean on one knee, that you can pick beans with two hands and beat all your buddies.

Paul picks alongside them sometimes to increase the challenge and demonstrate what's possible. But I am the labor foreman. The kids say I wouldn't last two days at General Motors—too strict! Wherever farm or business leaders get together they complain about the lack of good help. We don't. From the best of the bean pickers Paul picks his hourly work crew, and he'll proudly match them for work output against any group of $5- and $10-an-hour adults.

"Child labor!" It has a nasty ring left over from the sweatshops of the Industrial Revolution. America's children have been carefully insulated against the "cruelty of hard labor." I haven't met the kid yet upon whom you can inflict hard labor!

I have spent days and weeks and months riding herd on children in school "study" halls who were dying of an insidious boredom and were far more difficult to manage than my sweating bean pickers. The notion that young children cannot, should not, need not do meaningful work for at least a couple of hours every day is a ridiculous one. When I recognize the great value my sons place on their few free hours, I think that we must be lucky indeed.

Weather Musings

A FARMER LIVES WITH WEATHER AS A STOCK-broker lives with the market report. It is his first consideration in the morning, his last consideration at night, and sometimes his disturbing preoccupation throughout the night. When my grandfather was ten years removed from his last haycrop he would go to the window with deliberation, read the signs he saw in the clouds, and set up a fret about the storm front moving in. In the corner of our kitchen my father rigged up a rain gauge that he tended with as much concern as he lavished on the latest child. My husband on vacation in a far city arises at 6:00 A.M. and sits alone beside a TV set to get a morning weather report that is as useless to him as a pair of spats. Weather is a lifelong obsession with the country gentleman.

As a farmer's son learns where to find the draintile, when to breed a cow, how to rake a hayfield, which crop follows which in rotation, so he learns to read the weather in a thousand small signs. The moon, the wind, the clouds, the seasons, the caterpillars, the corn husks, the birds, the U.S.D.A., and the condition of his joints may all figure in his calculations. But he always has more knowledge of the subject than anyone but another farmer wants to hear.

The crucial weather reports of the year are the occasions of full moon in May and October. A still, cloudless night strikes terror in the heart of the fruit grower. Time was when all a man could do about the devastating frosts that nipped the buds and destroyed the early vegetable seedlings was wring his hands and pray. Even in the space age there isn't a great deal to be done, but technology has produced a few ideas to help the farmer sleep easier—or prevent him from sleeping altogether.

On the eve of calamity, if he hasn't already cuddled squash and cucumbers under paper hotcaps, he'll phone all his friends and the garden-supply stores in frantic search of some, and everyone will be out until moonrise applying the pesky things. It is really possible to hire a helicopter and a pilot to fly low over one's fruit trees and keep the air circulating, but failing that, a man can drive his air-blast sprayer through the orchard all night and accomplish a little more than a thorough chill. A gardener can run his irrigation during the dawn hours (to the irritation of the neighbors) and under a morning coating of ice miraculously salvage strawberry blossoms and tender vegetables.

But no matter what the efforts, morning sounds like Endsville. The naïve child who sits at breakfast and listens to the frightening talk of being "wiped out" carries a heavy burden of woe in his heart long after the adults have reassessed the damage, found it not so devastating, and failed to mention it. When another season of routine crops have materialized and catastrophe shows no signs of having struck, the child begins to wonder. . . .

But the next spring the talk is just as black, and again the child shoulders the burden, and again a crop comes along. By the third or fourth year of this "deception," he has learned a wonderful lesson in optimism—farmers' reports of crop damage are very much like weather reports: frequently unreliable.

Singing Spring

THE VERY LAST OF THE FARM JOBS THAT A farmer turns over to his hired help is the planting of his row crops. Someone else can fit the ground, cultivate, spray, harvest; someone else can keep the books and market the yield. But when the commitment of seed to soil is made, the farmer himself will make it.

He'll try to tell you that his reluctance to surrender this job is based on the fact that of all the variables involved in a crop, the "stand" is the most important. My private theory is that no one else can plant a row straight enough to suit him. A farmer is judged by his straight rows as he is judged by his stand, his clean fields, or his tall corn; a crooked row is a stab in his ego.

When a killdeer stakes a claim in the middle of a fitted plot and settles down to propagate her species, what's a farmer to do? Long before the ecologists started pleading the cause of the birds, his father and grandfather had taught him to treasure the killdeer. Clearly he has no choice but to curve around this creature so fiercely defensive of her eggs.

You have to admire the killdeer. No "rock-a-bye-baby" tree house for this mother, no protecting thorn bush, no leafy green shade, no cuddling warmth of soft grasses—only a barren circlet of pebbles in a rude field and a grim determination to survive and multiply. (A bird so average in size who produces four such large eggs is already heroic!)

Agriculture being what it is, this bird and her nest are no small disturbance to a farmer's rites. He must disc around her first of all and avoid her afterward with the herbicides. Then he must curve around with the planter and the cultivator, and watch that before the pesticides blow with their poisoning wind her babies have hatched and flown. He must reconcile himself to the fact that here, until the last gasp of summer, grows a patch of weeds. But worst of all, every time he passes that row he must defend that awful curve. No sign there that tells the neighboring farmer, "Bird's Nest!" Just a patch of weeds that summons an unfair judgment on his skill.

Ah, but written in the wind a message: "No, Rachel, no Silent Spring"—and next June again the clear, insistent call, "Kill-dee, kill-dee, kill-dee."

On the Critical List

WHEN I WRITE MY PRIMER FOR YOUNG FARMers' wives, I'm going to title one chapter "What to Do When the Combine Breaks Down."

The sun is shining, the wheat is ripe, the crew's on hand,

and the $%&#"%$ thing won't run! This is a circumstance so somber it casts a pall over the whole farm. It is to be treated with the seriousness of a death in the family, and requires many of the same actions and precautions.

In the first place, it's essential that everything else be subordinated to the "tragedy." The children must be immediately silenced and whisked out of sight if possible. They must be cautioned that their personal wants and desires are of absolutely no consequence for the duration of the emergency. Nobody or nothing detached from the immediate problem must be allowed to bug the farmer—farmhands, pets, customers, et al.

The preparation of food at such a time is of little importance, but as in the death situation it is a helpful diversion—for the farmer's wife. He won't care if he eats or not, but you can herd the children into the kitchen, fill their plates from the pans on the stove, and bustle about waiting on them, keeping conversation at a hushed minimum.

You can carry a plate to the shop where the farmer is working on the thing, but don't be forceful about suggesting that he eat. You may gently indicate that he "might feel better if he ate a few bites," but the better part of wisdom is to just set the plate down in a conspicuous spot as you would with a wounded animal. Let him make his own move.

The smartest thing to serve may be a cold beer. He's feeling overwhelmingly neglected by the powers of good at this point, and he longs to lash out in some act of "wickedness." For those who were conditioned early to the evils of drink (and most farm boys were), the cold beer is especially therapeutic. Besides, it's hotter than Hades and his temperature is ten degrees above that.

Much of what you do in the circumstances of a combine breakdown will be assigned by the "mortician" himself. For instance, "Take this hickey down to . . ." and from that point on pay strict attention. You must be as committed to orders as the leader of a commando raid.

Never mind that your jelly has just reached a full rolling

boil on the stove or that the baby is turning blue from a marble lodged in his throat. When you're standing in a farm implement store twenty-four miles away and the guy says, "Does he want a left-hand thread or a right-hand thread?" you sure as heck better be able to say something besides "Duhhhh. . . ."

You'll probably be doing a little speeding on the highway, and if a state patrolman stops you, just tell him the facts. "Officer, we've just had a death in the family."

But as is the case in all fatal situations, sympathy is your most appreciated offering. The "autopsy" will probably be taking place somewhere near the barn. Parts will be lying around in oily confusion. If the poor soul is struggling alone against fierce mechanical odds, he needs you, if for nothing else than to listen to him swear. Maybe you can hand him a wrench or hold a bolt, but be very cautious about making suggestions. Prayer sometimes helps, but not often.

One further precaution must be heeded by the zealous farm wife, especially the young and uninitiated: It is very easy in situations so tense and emotion-packed to get carried away and make concessions you may live to regret. Do *not* under any circumstances permit yourself to be persuaded that a new combine is the answer. A death in the family is one thing, but eternal damnation is something else!

Telephone Trauma

PAUL AND I ENJOY OUR LITTLE DIFFERENCES. There are a few standard conversations that never fail to set things off. Paul starts one popular number like this: "I'm going into town to talk with X (somebody)." This usually on the busiest day of the year.

"Why in heaven's name don't you just telephone him?" There follows a stony and reluctant silence. "That's just plain stupid! Driving all that way, using all that time and gasoline when you could accomplish the same thing on the phone." Lochinvar sets his lips tight, juts his jaw, and slams out the back door to his pickup truck.

There is something in a farmer that doesn't love a telephone! Perhaps he thinks of it everlastingly as a female contraption. I remember Grandma sneaking her phone off the hook to listen in on conversations of neighbors whose long and short rings she identified. And many a private line has been installed for people whose party lines were made impossible by yakking females.

There are, of course, routine uses for a phone that a farmer recognizes and accepts. He will check on machine parts by phone, place orders by phone, call the gas man or the vet or his breeding organization. He won't hesitate to call a few guys to play cards, and he finds that a telephone lends wonderful distance when he wants to lodge a loud and just complaint.

When he plods through the kitchen in his boots, lays his dusty gloves on the table, pushes back his cap, and sets himself reluctantly to one of these routine calls, a farmer feels he has an inviolate right to the line, and most women agree with him. A few gruff masculine words into the receiver result in hurried good-byes and clicks all down the line.

With the advent of the telephone recording device, a new dimension in frustration has been added to even the simple phoning he does indulge in. The farmer who has been swearing at machinery for most of his life shouldn't be expected to get uptight when confronted by an unseeing gadget on a phone, but not so. Trapped by the miserably compelling voice on the other end, he stiffens, swallows, and stammers, resembling no one so much as himself rising to express the dissenting opinion at the PTA.

There are certain items of business, however, that a farmer is definitely reluctant to transact by phone, and would *never* discuss with a tape recorder. Chief among these are money matters. If a man owes him money he will drive many miles, stand about dumbly, hem and haw, and finally broach the subject. Or, conversely, if he owes a man money, he refuses to inquire about the amount by phone. He prefers to take his checkbook, go, make his payment, and shake the man's hand.

He detests soliciting by phone for money or help or any other commodity. Only more than this does he hate to be solicited.

And my farmer certainly won't try to make a "deal" by phone if anything short of 300 miles of bad roads can prevent it. When a farmer deals, he wants to look the other guy in the eye, see and feel his responses. No amount of haranguing by a shrewish frau seems to alter that quirk one small fraction.

But I will keep trying . . . a good marriage thrives on little differences. And if I ever succeed in convincing him I'm right, maybe I'll be able to talk *myself* out of being terrified by telephones.

The Farmer Takes a Wife

ALONG WITH UNBELIEVABLE AMOUNTS OF money, every farmer must invest a few days yearly on the purchase of new equipment. Because a new combine or corn picker, bulk tank or tractor are often the sort of compensations she can look forward to instead of carpeting or appliances, the farmer's wife often tags along to ask a few questions, hopefully shrewd, always unwelcome.

Having agreed on one occasion that we needed a potato elevator, Paul and Ed (the farm manager) went off one afternoon to tilt with the elevator merchants. I wasn't invited to go along, and things had been piling up so I was content to apply myself to the mending. By and by, home came my true love with the facts: There were two elevators by rival manufacturers, one at $1300, one at $2000. What did I think?

I was thinking that with $1300 we could carpet the whole

house, and that with $2000 we could carpet the house and take a luxurious vacation! What I said was, "What does Ed think?" Ed thought that the $2000 elevator was engineered better. I reflected briefly that for $700 my father had studied four years of engineering. But, engineering or not, $700 is a healthy sum, so we tabled the matter for a few days.

Then came a trade journal advertising a used elevator of the cheaper variety, and a portion of one day was budgeted for examining it. This time I decided I'd go along and ask my stupid questions.

Lead-off question: "How has this thing worked out?"

"Well, we done a lot of work on it. We split it down the middle here and widened the track." (It sounded unbelievable; he indicated the repair.) "Ya' catch on?"

I didn't, but Paul was nodding and Ed chewed his gum.

"Had a little trouble with the boom. Wasn't long enough. We lengthened her and welded her here. Ya' catch on?"

Paul nodded and Ed chewed his gum.

"She leans a little," the fellow said evasively. "One of the troubles is the length of that chain. Too long. Needs an idler here. Ya' catch on?"

I looked at the spot but I didn't know what an idler was. Paul nodded and Ed took a closer look at that lean.

"The thing ain't made heavy enough. We come out here one morning and the whole business was collapsed on the floor. Had to weld an I beam to the frame here. Ya' catch on?"

Incredible! There was a stack of cement blocks loaded on the motor platform, and I questioned him about them.

"Not heavy enough on the back end. Needs more weight down here. Catch on? She cranks up good. Swings to the left good too, but she's a son-of-a-gun to swing back. I told them guys they was making that oscillator wrong. They fixed it on the newer ones. See? Down here. That hickey ain't right. Ya' catch on?"

Paul and Ed nodded and I started wishing I'd stayed with my mending.

"How much are you asking for it?" Ed asked.

"Well, I got a lot invested in that thing. I figure she's worth a thousand."

Meanwhile our children were walking on the boom, hanging on cables, climbing on tractors, pulling out light cords, and performing other attention-getting feats. The air was charged with suppressed discipline. I walked out, swatted them soundly on their seats, and took them back to the car.

The men talked for a few more minutes. Then I heard Paul say, "Well, thanks a million. We'll talk it over," and they joined me in the car. Ed smiled at me and said, "Did you catch on?"

The following day Paul went to the phone and I could tell that he hadn't just been nodding his head to mask his ignorance. He had caught on. He called the manufacturer and ordered the $2000 elevator.

I sat there wondering if it were possible to mend carpet warp.

Dad's Desk

I DO NOT PRETEND TO TRANSACT ANY BUSINESS at my husband's desk; I only depend upon it for a few staples like checks and stationery, stamps and envelopes. Further than that I lay claim to one drawer and half a pigeonhole. But I will say unequivocably that no desk is big enough for a husband and a wife.

A man is as possessive of his desk as of his razor, and nothing irritates him more than finding it a repository for the family junk. The man behind our desk has made it quite plain that motorcycle mufflers, bicycle seats, gym bags, and guinea pigs are not welcome thereon.

But what really throws him into a rage is an overflowing box of trading stamps (and when aren't they overflowing?). S&H stamps are more familiar to our children as G— D— stamps.

A farmer's desk harbors as much junk as the loft of his barn: seed catalogues from the year before last; three-year-old calendars; road maps enough to take him around the world; parts books for twenty-five pieces of machinery (Good grief! Did I call them junk?); and usually a drawerful of miscellaneous nuts, bolts, pipe fittings, and tools. My husband also keeps a book of jokes written in longhand by his father, and his grandfather's collection of Indian artifacts haphazardly filed in old wooden codfish boxes.

I don't really know what he has stashed away in the very dusty private realm of the filing shelves at the bottom, in twenty years I've never summoned the courage to investigate. He has a live filing cabinet beside the desk, but he's the bookkeeper and I don't meddle with that either.

Paul's desk also houses a lot of things common to desks everywhere: keys to vehicles sold years ago; keys to freezers and padlocks and suitcases never to be found when needed, keys to doors neither you nor your father nor your grandfather ever locked, forgotten doors to vanished buildings; miscellaneous foreign coins; dice; leftover 4¢ postcards; several 6¢ stamps, 5¢ stamps, 4¢ stamps, and two or three 8¢'ers without glue.

There will also be a small pack of foreign stamps saved for somebody's brother-in-law's collection, an old scratched magnifying glass, a pen holder saved for auld lang syne, lead pencils without lead, ballpoints without ink, a jammed stapler, a half-dozen marbles, and a magnet (why always a magnet?). The total impresses me as the lifetime overflow from a small boy's pockets. There are a few things one never really expects to find on Dad's desk: scissors that will cut (although plenty of the first-grader variety), cellophane tape that can be pulled from the roll without deft fingernail work, a pen or a pencil wholly equipped to write.

My husband's desk was originally my father's. It is a stately upright Victorian model with an open bookcase above—almost a handsome piece of furniture—probably the only reason my husband has for hating his father-in-law whom he never knew. As a desk it is cramped and inconvenient.

As far as Paul is concerned there is only one desk suitable for a farmer or any other man—a big blocky oak structure with multiple pigeonholes, broad work areas, deep spacious drawers, and a roll top to camouflage clutter.

His father had one and Paul would like to move it to our dining room, but his wife protests, "Honey, it's much too big for that corner, and besides it doesn't go with the other furniture."

He grumbles and sits, and starts sorting his way through the strata of debris, working his way down to the bills and bank statements. There's an aura of justifiable martyrdom about him. Just to tease a bit I inquire, "Did your mother share your father's desk?" I know full well what he's going to say, but he needs the satisfaction of saying it again: "*No* sir!"

I have a growing feeling that I'd better remeasure that corner, that someday soon interior decoration will yield to good sense.

The Country Husband

AT BREAKFAST THIS MORNING I REMARKED THAT the incidence of divorce seems less among the farm populace than among the population as a whole. "Why is that, do you suppose?" I ask Paul, and he has some opinions.

"A farmer can't afford to get divorced! And besides, there's something to be said for keeping a woman isolated in the country."

I suggested wryly that rural isolation hadn't been much of a deterrent for Lady Chatterley or Madame Bovary, but maybe he had a valid point on the economics of the situation. When a woman marries a farmer, however thorny their relationship, she grows to love his land more and more. . . .

I think about the subject all morning as I bend to pull weeds in the potato fields. About eleven o'clock—when my back seems permanently arched, my shoes are full of dirt, and my hands begin to blister—divorce for a farm wife takes on a certain appeal. Why is it that I always get stuck with the dirty work—the weed pulling, the hoeing, the supervision of the picking crews?

His cultivating done, Paul beats me back to the kitchen at noon. He has the potatoes cooking and is shaping hamburgers when I limp in. Moving numbly about, I reflect on the goodness and talents of my husband. A hundred hours a week he works without complaining, coping with frustration and hurdling obstacles over which I would weep in despair. A farmer today has no time to fritter on the end of a hoe—putter work for the less skilled (my classification).

But that is any country husband; this one is so much more. He can sew a pair of draperies as ably as he repairs a combine. He can iron a white shirt and wire a house. He can diaper a baby and repack a pump, do the laundry and overhaul the plumbing. He plays the piano and the organ, and when he must he can pick produce faster than anyone else on the farm.

He swims and skates and skis well. When the time allows he's a schoolteacher capable of teaching anything from phys ed to physics. He can cook and clean and hang wallpaper; he's a whiz on a typewriter, and he's quick to tell me when I have misplaced a semicolon or misspelled "phlox."

Divorce? Not on your tintype! Where could I find six men and a housekeeper to replace him? I'd better go wash my face. He says lunch is ready.

With Binder Twine and Baling Wire

RIDING DOWN THE POTATO FIELD BESIDE OUR potato harvester on this October day, I admire the complexities of this huge machine with its ten webs moving over rollers and sprockets and cones, interrelating through a system of meshing gears. Truly a machine in all its synchronic perfection is poetry in motion.

And then a clank. And a jerk. And a grinding of gears. The poetry dissembles into broken lines, obscure phrases, and single words—most of them profane.

Paul jumps down from his tractor and goes quickly, patiently about the tedious work of repair—the probing, the analysis, the selection of tools, the dissembling of the twisted wreckage, the instructions to bewildered assistants. Perching on top of an elevator and looking down at the tangled mass, I hug my knees and relax in resignation. I have been through emergencies a hundred times and one thing I know: My farmer can fix anything, and do it with dispatch. That is the genius of a good farmer. He knows exactly how to clobber a stubborn bolt, loosen a rusty nut, hammer out a crooked cotter pin. He knows just what part of his equipment can be utilized to aid the repair, how to improvise with a hunk of twine, a nail, or a twist of baling wire.

During the Marshall Plan years, depressing and scandalous reports were made of nearly new machinery abandoned in fields of want all over the world. The greatest thing wanting in those sad situations was good old Yankee know-how and make-do. This is a talent bred in the bone of the farm boy, starting in the sandpile with a broken hitch on a toy trailer, proceeding through tricycle pedals to bicycle sprockets to tail pipes on jalopy trucks.

The world is full of mechanical specialists, but for all-purpose use give me a farmer, a rustic wizard who can take a string of loose profanity and hammer it back into "poetry."

"A *Bold Peasantry Her Country's Pride* . . ."

PAUL IS UP IN THE NIGHT AND GONE TO market at the Cleveland Food Terminal. It is the weekly scheme of things.

"Does Daddy like to go to market?" asks Orrin. It's a question that probably never occurred to Paul. A little boy listening around a supper table long ago to his father's chatter of the market, to talk of customers and prices and quality of produce, grew up to the expectation of taking his father's place at the market. Now he assumes that responsibility with the same sort of obedience and devotion he brings to all his farming.

I have lately read a classic novel of Norway, *Growth of the Soil* by Knut Hamsun, a moving and timeless portrait of the peasant. Isak was a man who took on his back his basic needs; went into the wilderness, chose a spot from the quality of the molds and the rotting verdure, for the presence of wild game and the availability of the water; then settled to build a life from scratch.

He was a brute of a man, "a brother to the ox," unlettered and unrefined. But he created a small farming empire through the application of hard work to the understandings of the peasant. Edwin Markham might have written of him, as he did of "The Man with the Hoe" (Millet's painting of the French peasant), "Who made him dead to rapture and despair . . . What to him are Plato and the swing of Pleiades?"

But I have a long quarrel with Markham's word portrait of the peasant. If he had looked at more of Millet's paintings, he might have discovered other sides of the peasant than the evening backache. He could have balanced it with the night's deep untroubled sleep. What, indeed, are Plato and the swing of Pleiades to the farmer who hears the song of the lark at dawn?

You marry a man and you live with him and you think you know him—as I with my farmer. Then along comes a prototype in literature like this man Isak and you realize how little you have understood. (I wish I could make this point to students in literature classes when they ask, "Why do we have to read all this junk?") I am like Orrin asking myself, "Does he enjoy his work?" Anybody can love to farm on a bright spring morning, but there's so much more. How about plowing in the late March chill, cultivating in the summer heat, standing in a market lot in a November rain?

Paul doesn't waste time asking himself questions. He chose this life like his father before him, and his father before him, and his father before him. Who knows how many generations of peasants brought him to the fullness of his instincts. His very dedication makes the question of enjoyment irrelevant. This work, this land, this life transcends pleasure for him. It is complete fulfillment.

Isak had a wife, Inger, chosen at random because she filled the need of someone to care for the stock and help with the harvest. His love for her grew from her filling of his needs, first as a farmer, second as a man.

Inger wanted to be part of the world, know more than her remote life in the wilderness. Through strange circumstances she gained her chance. She added knowledge and refinements; she broadened her experience and acquaintances and grew discontent. But it was the solidity, the simplicity of Isak that saved her, restored her balance and perspective.

We, Pat and Paul, are like that. In our basic natures we are Isak and Inger. Paul, of course, has many refinements that Isak lacked. In contemporary jargon you might call him a "country gentleman," but scratch a country gentleman and you find a peasant.

I will have the music and the dreams, Plato and the

Pleiades; I will chase rainbows and butterflies and ideologies. But Paul will have the forest and the field, the majesty and power; he will know the sequence and the purpose of things. He will make things work and put food on the table. He will do while I dream.

The boys talk of the future and the things they will be. Orrin thinks maybe he'll be a doctor "because they make lots of money." And I would turn to him and say, as someone does to Isak's son in this beautiful book, " 'Tis not money the country wants, there's more than enough of it already; 'tis men like your father there's not enough of."

3 CHILDREN

. . . trailing clouds of glory do we come
From God, who is our home

William Wordsworth, ODE ON INTIMATIONS OF IMMORTALITY

It was the day that Teddy at age four stood on my clean white tennis shoes, clasped my hands in his, and looked up into my face with joy saying, "Somebody digged this all up for me!" that I knew I wanted to write. He was talking about the bare spot in the back yard that I had laboriously spaded and raked and seeded to grass and that he had reduced in a few minutes to a glorious sandpile.

That is surely what it is to live with children, to watch them transform your own best efforts into something undreamed of, uniquely their own. It is to learn patience in spite of yourself, to live tenuously between trauma and delight, to discover truth at its source.

Age of Wisdom

A FOUR-YEAR-OLD RIDING BACK AND FORTH AS an incidental passenger in a kindergarten car pool is bound to absorb a great deal of bewildering information. Once during his involuntary year of commuting, Orrin registered a premature but very just protest against regimentation.

There had been a discussion among my kindergarten passengers of a culprit who had put slag in the toilets and been reported by fifteen slightly less guilty little boys.

"Do you mean that in kindergarten everybody has to go to the bathroom at the same time?" he asked out of a clear blue sky several hours after the conversation.

Obviously he had given this ridiculous idea a great deal of thought. I was somewhat flustered, and there followed a silent interval while I collected my thoughts. A mother's answers must help a child to make sense of an often senseless world. I thought at first to explain that whether or not you had to "go," you still made a pilgrimage to the restroom with the rest of the class. But this was just to admit to the idiocy of the system and the adults who create it. Better to let him think that like Pavlov's dogs, schoolchildren can be conditioned to go to the bathroom at the same time.

"Yes, I guess they do," I answered.

Turning back to his play, and in a "whattaya—take-me-for" tone of voice, he said, "I don't believe that."

I stood there pondering defeat. Something had suddenly become very clear to me. I understood why that little boy put slag in the toilets!

Tale of Two Kites

THE VIEW FROM THE FRONT WINDOW ONTO THE playground across the lane is of two kites anchored by three boys, and no one wonders at the miracle more than I. Paul and Ed walk off the field and down the road, leaving the enterprise to my sons. As I go out of the kitchen to make my admiration known, Ed smiles and hollers to me, "First successful thing I've done today!"

The kites are very high and sport long tails, which also surprises me; no one has troubled me for rags or coaxed me to be a party to the ripping.

"Where'd you get the tails?" I call to Paul.

"Ed and I made them out of the red-and-white tablecloth."

That tablecloth was a part of our "dowry." For five years I'd thought of throwing it out, faded and ragged as it was, but sentiment runs deeper than pride so I ironed it time and again. Finally last week I folded it lovingly and put it in the rag drawer. Now it ascends to glory on the tail of a kite.

I watch the maneuvers a while longer. Teddy's kite flies

higher than Orrin's, and I feel my soul soaring with it. When the kite dips drastically I catch my breath. Down, down, down; the tail drags the ground! Then an updraft takes it, and somehow I know my Teddy is lifted on the tail of that kite out of the frustrations that sometimes plague him.

Dane helps his smaller brother, and after five minutes of just standing there holding the string, feels that some adjustment should be made. Nothing duller than a kite that flies precisely as a kite should. He pulls it in, and Orrin runs along after him as they carry it into the barn.

In fifteen minutes it's all over. Maybe they'll sail again, those kites, but excitement will tempt them close to the trees another day and that will end kites for this year. A good half hour must be the yearly average on one kite for one boy, but it's a thrilling, exhilarating half hour that lends as much enchantment to a man's memory of childhood as many an hour invested in greater pursuits.

If that boy is normally resourceful, however, he will discover that the raw material of a broken kite is precisely what is needed to make a very satisfactory bow and arrow, a metamorphosis that very happily extends the tale of a kite.

In Search of Spring

It was a gray, depressing day with a biting wind; the only hint of spring was a barely perceptible green over the rye field. I didn't need to remind anyone to take his gloves or zip his hood, though my protestations that a trip to the river demanded boots were met with the usual scorn. So off they went, three eleven-year-olds in dirty tennis shoes and Saturday blue jeans, chewing on apples snatched up as an afterthought from a basket at the back door.

Teddy was bare-headed and had a swagger in his stride that identified him as resident "country mouse" leading two city cousins. I watched them from the kitchen window until they sank from sight over the brow of the hill, and then my thoughts followed them through the afternoon.

When little boys go off to the woods, it's surely a private excursion. But if you've grown up in the country you can almost chart the adventure. First you kick up a good strong stick for cracking against trees or lashing at the bushes. And if you find an animal hole, you poke at it with that stick and wonder if there's a fox in there.

In a wet bog you come upon an emerging skunk cabbage, but with two other fellows there you don't make much of it. With your toe you kick at the swirling purple, then scuff on.

Streams trickling out of the hillside are more justifiably fascinating. In their secluded gullies the snow usually lingers and you poke at that with your stick. There's a special pleasure in kneeling and drinking from one of these clear icy streams, especially if you intercept it at a small waterfall. Or perhaps you break off a late-hanging icicle there, sucking at it awhile and then dashing it against a rock.

There are many paths to be taken to the river through the woods, but having discovered a gully, it's inevitable that you follow it, climbing downward over rocks and fallen logs, the underbrush lashing at your face and catching on your clothes. This kind of trekking is much better done with high boots, but from the tennis shoes that I often discover cemented with mud to my cellar floor, I conclude that the discovery is the same no matter the footgear.

Rivers offer their own diversions. You find a bottle or a tin can and fling it in, hoping it will float so you can throw rocks at it. And, of course, you search for flat rocks and skip a few of those across the muddy surface. The river is usually too high and fast-flowing in March to cross without a boat or a raft. After little boys have run along its banks chasing floating objects until the enthusiasm palls and their feet are good and wet, they start dreaming up a raft.

Now is the time to throw in your stick, signaling an end to the present adventure, and turn homeward, talking excitedly of the craft that will bring you back (on a warmer day it is hoped) and carry you downstream on another afternoon's adventure.

I didn't see the boys return. I only knew they were back when I went out to shake rugs and saw a cloud of hay chaff poof suddenly from the upstairs door of the hay barn. The sun had come out, the rye looked suddenly greener, and the breeze had taken on an April warmth. It was then I realized that the boys had really gone in search of Spring. And lo and behold, they brought her back!

Flower Power and the Generation Gap

MY MIND, AS EVERYONE'S, OVERFLOWS THESE days with troubled thoughts of youth—rebellious, undisciplined, defiant. What's to become of them? Of us? Of all that the generations have built? Why do we have this conflict between us?

But you can't isolate yourself in troubled thought with May bursting around you. The leafing trees, the bird song, the flowers will not be ignored; and loudest of all call the poppies from the garden by the clothesline. I bring in a few furry buds, singe their crooked stems, and watch them as I go about the household routine.

And so it is that I discover . . . a poppy does not unfold slowly from its calyx as most flowers do, but is held prisoner by it. Then pushing, pushing, pushing from its center, it ruptures the calyx from the stem and in a rebellious burst of opening, shucks and drops it. Once free of the calyx, this becomes a full-blown flower within an hour (here out of the wind it takes longer).

And then that cry of "Look," that splash of orange, that flamboyant day of life! One beautiful day—ebullient, joyous, unrestrained. Then a day of sagging dotage, and next morning the orange petals are a mound of regret on the table.

I say "regret," but how can I know what a day of life

swaying in the wind might be to a poppy? One vibrant, all-glorious day might be an infinity of joy. Yet certainly it does not satisfy me. Anything so lovely should not exhaust itself so soon. Since April the tulips have stood stubbornly against the wind, changing in hue and aging, supported by a calyx that remains part of them.

The problem with the poppy, surely, is with the calyx! Why does it resist; why doesn't it give a little, stand back and let the flower emerge? Why does it first repress the bloom, then desert it and let it go into the wind wanton and alone?

A day ends and I am through with coming and going, with "poppy watching," too. I put my little children to bed, rinse the bathtub, and pick up the clothes, stopping a few minutes to sit beside my teenager and exchange evening thoughts. He listens to my problems and offers some good suggestions, then turns the conversation to the things that concern him.

His thoughts are not my thoughts, his dreams not my dreams. Nobody ever promised that they would be. But suddenly I want to foist upon him all that I value and have found satisfying, point out the shortcomings, the folly of his dreams. Some repressive urge in me would stifle them, change them.

But downstairs on the table a poppy calyx lies withered and dry, cast off, useless, as a flower blooms too quickly and too briefly.

Fish Story

When I dreamed the childhood dreams about "my children," they were always towheaded girls who would sit on porches on summer days playing house with their dolls. Somewhere in the attic are some thirty-year-old dolls packed away by a disillusioned dreamer, and I'm wishing I'd paid more attention to my brothers at their fishing.

I remember an afternoon when they lent me a line with a hook and a cork and I caught twenty or thirty fish. I've never told my kids about it. They wouldn't believe me. "You mean to tell me you caught thirty fish with nothing but a line and a cork, a steel nut, and a hook with a worm! You're too old to remember straight!"

It's certainly not that simple with my sons. Strewn all over the house, the basement, and the back porch is fishing gear: spinners, plugs, sinkers, bobbers, flies, hooks, nets, lines, reels, rods, stringers, tackle boxes. I find myself wandering helplessly through the sporting goods department, asking a hapless clerk, "If you were a black bass, which one of those 300 lures would you be attracted by?" He picks a red popper with an Indian feather in its brow, but I don't trust him and come home with a silvery, minnowy-looking creature as well.

It turns out we were both wrong. "You know what black bass bite best on, Mom? Purple plastic worms," says Teddy.

Well, who'd a' knowed it? (Did you ever grope your way around in the night and step on a purple plastic worm in your bare feet?)

Gosh, remember the guileless barefoot boy in patched blue denim who used to meander down the lane on a lazy afternoon carrying a can of worms and a fish pole cut off a willow tree? His object was a fish—any old fish—in a muddy brook.

Well, the barefoot boy in patched denim lives. But, believe me, nothing else is the same! These guys bound out of bed at 6:00 A.M., look at the sky, say, "Looks like a good morning for shiners." Then they grab rods and reels and the rest of the paraphernalia and head for a scientifically stocked pond. The worst of it is, they sometimes catch fish!

And what does a squeamish housewife, harboring quiet dreams of little girls and dolls, do with a fourteen-inch bass gasping in a bucket on the back porch at seven in the morning?

On mornings like these I think back in sympathy to an old college friend struggling to rear five children in a modest suburb, who blurted to me almost in tears, "It's really shattering to discover that a college education hasn't prepared one to cope with any of the practical realities!"

Rumble at Rugby Corners

RUGBY CORNERS IS THE MOST UNLIKELY PLACE in the world for a rumble, but on Saturday there's going to be a gang war there. Like most wars, this one started with two individuals. Leroy and Greg had a quarrel over a camping tent or something equally vital to boys of twelve or thirteen. The quarrel festered and spread. Greg had a friend and Leroy had a friend, and the friend had a brother or a sister. Sides were drawn. Organizations formed to write up grievances. Some families were split for romantic reasons. Peggy had a crush on Greg and joined his club, but her brother was a friend of Leroy. The perennial enthusiasm for slingshots flared, and the idea of a war evolved. I was innocently drawn into the conflict as a munitions supplier. My navy beans were confiscated for slingshot practice.

Concurrent with the war buildup, I was organizing a fund campaign for American Field Service, a foreign student organization whose motto is "peace through understanding." Noting that a lot of small-fry energy was going to waste in Rugby, I suggested to Orrin that his club members might help me canvass for A.F.S. on Saturday.

"No, we can't help Saturday. That's the day we're having the war," he said.

It was then that I learned all the particulars of the conflict: that the battle was to take place in the yard of the old red schoolhouse, who was on which side, the specifics of weapons escalation, etc. . . .

"Sounds to me as though your side is outnumbered," I said.

"Yeah, we are," he said. "But we've got better weapons and ammunition than those guys. We've got membership cards, and a chemistry set, and one of those things you look through." He pantomimed a microscope. Obviously they were organized, and armed for chemical warfare. "Those guys are really scared. Leroy's planning to ask some sixteen-year-old kid. Greg says if he does that we're gonna' get some big buys too!" (He went on to name a few of the high-school football players.) Needless to say, Mother began at that point to throw a monkey wrench into the war machinery.

When Orrin was five and had been shooed from the kitchen one evening so Mother could "have some peace," he came to me wistfully, asking, "What do you do when you have peace?"

Today I came to him with a suggestion of what he could do Saturday in the cause of peace. But it's too late. He has a war planned for that day.

Yes, Rugby Corners is about the most unlikely place in the world for a war—unless you think about the bridge at Concord, the courthouse at Appomattox, San Juan Hill, the island of Iwo Jima, or the quiet jungles of Phnom Penh. . . .

Letter to a Son on Mother's Day

ONCE I LAY IN LABOR CLUTCHING FRIENDLY hands and breathing deeply and yearning for an end; and when the fog lifted they lay a small form in my arms and it was Mother's Day. A hundred nights I sat upright and asked, "Is he breathing?" and ran to a cradle and touched your warm body, and it was Mother's Day.

On sunny afternoons you stood in your crib and reached for a sunbeam—Mother's Day. You grasped for tangible things, baubles I cherished that broke at your touch, and I wept and knew Mother's Day.

I have heard your screams and run with a towel to stop

your flowing blood, and sat in emergency rooms stroking your head and holding your hot hand and praying on Mother's Day.

I have looked at group pictures of disheveled children and seen only one face—yours. I have sat in auditoriums where out of 100 performing children only you stood out.

I have watched you at play with strange children when you stood aside shy and frightened; and I have watched you lead a charge on the haymow with all the neighborhood in pursuit. I have seen you unite with them in frantic projects, and then gone at a tug of your hand to inspect tree houses, tent houses, caves, leaf piles, snow forts, and snowmen—always it was Mother's Day.

I have scolded and chastised and paddled; cajoled, laughed, applauded, advised. I have untangled fish lines, tied tails for kites, sewed marble bags, laced ice skates. I have made milk shakes, baked cookies, packed picnics—performed the many joys for you that complicate and enrich a mother's day.

I have screamed at you in my impatience over unimportant things, and gone to you in your hurt and apologized. I have recoiled at words or deeds "good" children do not tender toward their mothers, and been overwhelmed with forgiveness when you came to say, "I'm sorry." It was a mother's day.

I have shared you with your father and delighted in the sharing. There were times of arbitration when I explained you to each other, and times when you drew apart for father's days I could not share.

I have given you over to other mentors—to teachers, ministers, 4-H leaders, coaches—and been grateful to them for what they gave you of themselves, proud to share my mother's days. But I have been jealous that always you gave them your best face while at home you bared them all, which is as it should be.

You have come clattering into the house with the smell of schoolrooms heavy upon you and shouted, "Mama!

Guess what?" And I have guessed a thousand times, and known that the only truth was you wanted me there—Mother's Day.

From the window I have watched you at play and at work, developing strength and independence, and I have felt the tug of the "silver cord." The world claims you more and more, and I go to bed not always knowing where you are, but trusting you and loving you always.

You bring me a plant for Mother's Day, or a handmade card, a paperweight, or pincushion, or you bring me nothing more than you have already given, and certainly you need not. I gave you life, and every fulfilling day since, you have given me back something wonderful of yourself on a succession of endless Mother's Days.

Battle Campaign

MANY CHILDREN LIVE IN SECLUSION BEHIND walls of their own construction. You may peer through and agonize for the loneliness there, but you enter only upon invitation. In precious moments when no one can see, a small hand is slipped into your hand and you know there's a break in the wall.

I do not live well with walls. Communication through them is strained, so Teddy and I have our "battle encounters." This one started with muddy shoes and climaxed in a slap. In a silent surge of temper I rushed him through the bath ritual and into bed. Instead of hearing prayers I was invited to "Get out!" The wall was armed and the guns were firing.

Wounded and unhappy, I fled to my basement laundry, where my hands went automatically about familiar tasks. Mentally I reviewed the battle and angrily justified my position. Deeper than psychology, deeper than maternal love, deeper than understanding runs my father's edict, "Children don't speak to their parents like that." Justice was on my side! But justice doesn't cure the ache nor tear down the wall. Justice is no substitute for compassion.

A skinny child in striped pajamas appeared quietly on the cellar stairs. "Mama, I can't get to sleep." This was truce. And over a storybook in the corner of the davenport, we hammered out a peace.

As we climbed the stairs to bed in that rare moment of solidarity, he put his arm around my waist and I hugged his thin shoulders. Beside his bed I heard his prayers at last.

"Would you stay by me awhile, Mama?" he said, tightening his hold on my hand. Still on my knees, I lay my head beside his. There within his "wall" I dropped a white flag as we both fell asleep.

Bum Steer

"THAT," SAID MY FRIEND ELEANOR, WAVING her hand at a fold of 125 sheep, "is a 4-H project that got out of hand!" Ah yes, I thought, but at least that flock took a few years. The year we struggled with 4-H steers they were out of hand from the very beginning!

Clyde and Charley were Hereford calves Dane fattened the year he was fifteen. I admired their square brown bodies and their curly heads as we bedded them down for a nourishing year in the old horse barn, little suspecting the adventures we would be heir to.

When a sobbing boy burst into our bedroom before dawn the next morning crying, "Dad, my steers are gone!" 4-H took on new meaning: Hurry, Hunt, Help, and a brief expletive muttered by Dad.

Out we all went in boots and pajamas. It was our most memorable chase, but certainly not our last. "Charley slipped his halter!" became a rallying cry that brought everyone to their stations like volunteer firemen. The neighbors chased, the mailman chased, the school bus driver chased. And on the day we hosted the church picnic, everybody including the preacher chased.

Clyde and Charley grew in bulk, despite the exercise, while the Leimbachs developed the lean look of trackmen.

Like most teenagers with wheels in their heads, Dane had happy visions of the Honda he would buy with cattle profits. Two steers bought for $270, gently tended and nicely fattened, might bring $600 at a successful fair auction—a profit of $165 each. Wow! First feed bill for $25 brought the dawn of reality.

"Gee, Mom, it's gonna cost me $300 just for feed, and I've got hay to pay for besides that!" It became evident that in a year, two steers could easily eat a Honda. Charley got a runny nose, and a vet bill was subtracted from diminishing profits. In the feeding and grooming process, five buckets were destroyed ($4.25). Dane needed two rope halters (rope, $2) and a TB test (veterinarian $7). Fiscal fiasco!

The grooming of a 4-H steer for show is a process that makes men out of boys and heroes out of fathers. In those weeks before the county fair, a steer must be trained to walk docilely around a show ring, whisking his combed tail flirtatiously, stopping at the tug of a halter, lifting his head and setting his feet to show his deep brisket and his broad rump to advantage, and otherwise impressing the judges that he is a prime beast.

Charley was obliging as Dane trimmed his hooves and curried his rump. He displayed his conformation as though it were his glory and not his doom. Clyde had other ideas about grooming, and it fell to Dad to teach him better ways. How thrilling for a wife to watch a steer dash up the lane towing her true love by the seat of his Bermuda shorts. Watching her spouse turn a couple of somersaults and arise bruised, bleeding, but still in command of the animal and his halter, confirms in middle age her early conviction that he is of heroic caliber.

Three times in one evening an angry Paul dragged a mutinous animal back to the training ring with brute force. The word "bullheaded" suddenly became graphic ("Gee, Dad," I muttered, "where were you when I was trying to train the kids?")

When fair week finally came, a wary investor lay down in the Junior Fair barn beside his four-legged investment. Would Dane lead Clyde, or would Clyde lead him? Would the sale price cover expenses, or was the whole year's work and worry gone for naught?

Grandma and our suburban relatives came by to wish Dane well. "Beautiful animals," they said, but what did they know? Truly, Charley was beautiful, but Clyde was obviously built for speed. One wisecracker suggested that if he didn't make it in the steer class, we might put him in the trotting races. Cattlemen walking by would remark, "There's a lot of daylight under that animal." His back sagged, and no amount of currying would conceal his ribs.

On show day Charley, scrubbed and combed, was set aside as one of the top three contenders in the ring. One promising animal was being shown expertly by a shapely girl with flowing blonde hair. While Charley's trainer stood transfixed by the blonde, Charley relaxed like a matron let out of her girdle, and was dropped to fifth place. Dad tried frantically to signal Dane.

Clyde was on his good behavior; he ran Dane briefly about the ring, and fell in line just as Dad prepared to vault the fence and take charge. But soon Clyde was relegated to the lower end of the line with others who "also ran."

The auction was an inadvertent success. The comely blonde trainer of the Grand Champion inspired the buyers as well as the 4-H boys. Charley, as No. 5 animal, brought a price that balanced Clyde's slimmer pickings. But even then the total only covered Dane's feed debt. When the steer check came in the mail, he handed it over glumly. "Here you are, Dad. I guess it's all yours for the feed and hay."

He had lived with the knowledge that there would be no profit, but this was surrender and it hurt.

"Well, Dane, you've learned what we all learn sooner or later, that investments don't always pay a dividend. The margin in the cattle market can be like Clyde—thin. I've fed a lot of cattle, and some years I've given away my hay and corn. What Clyde and Charley ate won't make me a poor man, so you just give me what you paid for the calves and we'll call it square."

Dane's face brightened; he was both relieved and delighted, but he wasn't about to blow his cool. "Gee, Dad, you're all heart!"

"Yup," said Dad. "I'm 4-H clear through—Heart, Head, Health, and Hand-in-the-pocketbook!"

Fair Pigs: Notes and Comments

TEDDY'S 4-H PROJECT THIS YEAR IS MARKET pigs, and it's high time somebody started showing an interest in readying them for the fair. Studying a publication called, "Training, Grooming, and Showing Guide for Swine" I conclude that there is more light yet to break forth in the making of a 4-H mother:

"ITEM 1: THE PIG SHOULD BE GENTLED BY HANDLING WITHOUT BECOMING A PET. . . ."

The first question that occurs to me is, how do you make an "unpet" of a pig? This handy-dandy grooming guide comes into our hands two weeks before the fair, when Wilbur the pig is standing on his hind feet licking his master's face.

Reading on, I begin to see how one might accomplish this: "IT SHOULD BE TRAINED TO RESPOND TO EITHER THE CANE OR THE WHIP. . . ." That sounds like a splendid way to destroy a close relationship.

"ITEM 2: FIRST GET THE ANIMAL TO LIE ON HIS SIDE . . . BY MERELY STROKING HIS BELLY." (Don't make a pet of the animal.) "THEN TRIM THE TOES TO THE PROPER LENGTH. . . ." Sounds like a neat trick to a woman who never trimmed kids' toenails because she couldn't strap them down! "ALSO

SHORTEN AND DRESS DOWN THE DEW CLAWS." (Don't forget the dew claws . . . or the "don't" clause: "Don't make a pet of the animal.")

"ITEM 3: TUSKS IN BOARS ARE DANGEROUS FROM THE STANDPOINT OF THE HANDLERS AND OTHER ANIMALS." It depends, of course, on how close you're standing and how fast you can run. Number the handler's mother among the "other animals."

"TO REMOVE TUSKS, TIE THE ANIMAL TO A POST WITH A STRONG ROPE, DRAW UP THE UPPER JAW AND USE A BOLT CLIPPER." And if the pig gets loose, run like (blip)!

"NOSE RINGS MAKE THE NOSE SORE AND THE ANIMAL IRRITABLE AND UNMANAGEABLE." Which means that pig and handler will probably be of one temperament. "REMOVE RINGS . . . WITH WIRE PLIERS OR NIPPERS. . . . " And if that nose ring was serving a good purpose, maybe that old pig will dig you a tunnel clean through to the center of the race track.

"ITEM 4: CLIPPING . . . TO ENHANCE THE TRIMNESS OF THE ANIMAL, AND TO SHOW OFF THE TEATS OF A GILT . . ." Sounds like an ancient Egyptian seduction rite. (It should be noted that the handler hasn't had a trim for eight weeks.) I wonder whose razor is to be utilized for shaving this pig. Dad goes into shock if he discovers that someone has used his to shave a hair off a wart, and I'm surely not going to volunteer mine! We have relatives with poodle clippers; could you clip a pig with poodle clippers? We don't have to tell them what we are using them for.

"CLIPPING INCLUDES . . . EARS, BOTH INSIDE AND OUTSIDE; TAIL, EXTENDING FROM ABOVE THE SWITCH TO THE TAIL HEAD; THE HEAD AND JOWL: THE BELLY OF GILTS." "Hey, Dad, is Wilbur a gilt?" What, pray tell, is the pig doing while all this clipping is going on? (Don't make a pet of the animal!)

"ITEM 5: WASHING . . . RUB THE HAIR WITH TAR SOAP . . . ADD A SMALL AMOUNT OF BLUEING TO THE WATER WHEN WASHING WHITE HOGS . . ." (Clorox needed here, and here, and here). "FOLLOWING WASHING, PLACE THE ANIMAL IN A CLEAN PEN TO DRY." Another neat trick. Ha ha ha!

"ITEM 6: OILING COLORED HOGS . . . TO GIVE THE NECESSARY BLOOM TO THE COAT." Blooming pig . . .

"ITEM 7: POWDERING WHITE PIGS . . ." They've gotta' be kidding!

"ITEM 8: SHOWING . . . IN ORDER TO WIN . . ." The nitty-gritty. "A. TRAIN THE ANIMAL LONG BEFORE ENTERING THE RING." Well, it's too late to make it on that score. "B. HAVE THE ANIMAL CAREFULLY GROOMED AND READY FOR THE PARADE BEFORE THE JUDGE." Why else am I wasting my time on this mammal? "C. DRESS NEATLY . . ." So they'll know you from the pig. "D. ENTER THE RING PROMPTLY WHEN THE CLASS IS CALLED. E. DO NOT CROWD THE JUDGE." Judges are very testy about being crowded, especially by pigs. "F. WORK IN CLOSE PARTNERSHIP WITH THE ANIMAL." Us pigs have got to stick together. "G. BE COURTEOUS AND RESPECT THE RIGHTS OF OTHER CONTESTANTS." Do unto other pigs as you would have them do unto you. "H. DO NOT ALLOW THE HOG TO BITE OR FIGHT WITH OTHER ANIMALS." If there's to be peace in the world, it's got to start with pigs. "I. BE A GOOD SPORT. WIN WITHOUT BRAGGING AND LOSE WITHOUT SQUEALING." And keep that damned pig quiet too! "J. KEEP CALM, CONFIDENT, AND COLLECTED" (Neat trick No. 3). "REMEMBER THAT A NERVOUS SHOWMAN CREATES AN UNFAVORABLE IMPRESSION." Especially when he's being trampled by forty pigs!

Lamb to the Slaughter

I WONDER IF ANYONE WHO HASN'T LIVED through the experience of watching his child show an animal at the county fair can begin to appreciate the drama of the situation. I sit up here in the stands overwhelmed by feelings of helplessness and pride. I've contributed all that it was my part to contribute, which wasn't much. I brought a clean shirt for the showman and combed his blond hair while he tied his tie. My face wears an "isn't he cute" sort of grin that I'm sure Orrin would not appreciate if he weren't so intensely involved in what he's doing as to be unaware.

What's concerning him is the unpredictable actions of the lamb he's clenching in his sweating hands. He is concerned too with the judge, and his animal's relationship to that most important gentleman. The children have been coached to keep their animals between themselves and the judge in the best stance possible. As the judge moves about the ring, the boys and girls rotate themselves and their animal toward him as in some strange ballet.

The atmosphere is one of great intensity as the judge taps one animal and then another, instructing the young shepherd to move him into an elimination sequence.

Some animals are docile and obedient, others lively and difficult. One small, dark-haired boy with an unruly lamb bears a look of constant distress; as the animal suddenly breaks away and dashes out of reach, his young face cracks in tears. A wave of sympathy moves toward him from the crowd and the other competitors, and some of the older boys succeed in trapping the culprit. With more strength than he thought he could command, the little boy moves the lamb back into position and kneels beside him.

Orrin is whispering calming words in his lamb's ear; I think with pain of the time two days hence when Christian will be sold at auction and Orrin must part with what has become his pet. It wasn't intended that way; he assured me at the outset that he would try to remain detached. But how can you not love a lamb who follows you about, begging for affection as well as food? There was something symbolic in the name; when was it that a Christian wasn't born to sacrifice?

The judge succeeds finally in ranking the animals. Orrin and Christian hang on till almost the last, the judge feeling up and down the wooly back for the depth of the loin, grasping the leg for the quality of the muscle, and finally waving him into the second row.

One has to feel great sympathy for the boys and girls whose animals bring up the rear, having been judged too light or too unfinished to make the cut. So many hours of tedious chores have compounded to build this moment, so much of rolling out of bed early to feed the critter, of sacrificing outings or whole vacations because there was no one else to care for it. There was the nuisance of keeping record books, of badgering someone to stop at the mill for pellets or a salt block. There was the expense of paying a veterinarian to wave a blessing over him, and the bother of carting him to the fairgrounds earlier in summer for shearing, not to mention the arrangements that were involved in getting him to the fair and caring for him here.

Now some expert taps him with a cane and judges him second or third rate. Really hard to take. And the parent who watches it happening suffers twice—for the child and for himself.

It occurs to me as the scene is played out that the total experience of these 4-H children with these animals is the essence of the life experience: to commit yourself to something (somebody, some task even), to care for and struggle with it, to develop in the process a love for that object, to be made accountable for what you've achieved with it, and then to give it up and move on to new commitments, new involvements. This is to grow and to mature in the solidest way.

Dozens of Cousins

LIKE MOTEL AND HOTEL PROPRIETORS, WE have at summer's end an assortment of possessions that guests have forgotten—odd socks, billfolds, worn tennis shoes, underwear with unfamiliar labels, sweatshirts emblazoned with the names of rival schools, jeans with holes in the knees, and toothbrushes with hostile bacteria—left chiefly by varied species of the genus "cousin." A summer of twelve weeks has brought eighteen weeks of cousins, which means that often they were three in a bed or two on the floor, or stacked up in heaps in the corners.

Despite Uncle Paul's protestations that *"This is not a fun farm!"* his failure to utter a definitive *"No!"* (Aunt Pat, of course, is a pushover) usually brings out of hiding behind some kid's back a brown paper bag with church clothes, a toothbrush, and a change of socks and underwear. Then there are cheers all around as he climbs into the station wagon to begin his turn on the farm.

I have tried to analyze the appeal of the place. It's certainly not the food. A kid who can't rustle his own breakfast and at least a light lunch bids fair to starve at End O' Way (but then even cooking is a freedom not every home allows). The housing is crowded and uncomfortable—six or seven or eight to a single bathroom is prehistoric to the split-level generation, but three in a bed is as much fun as it ever was.

The schedule is erratic. Aunt Pat in her disorder frequently fails to notice that it's 9:30 and nobody's even thinking about bed. Neither does she feel constrained to make you eat your vegetables or drink your milk. If you don't make your bed in the morning (if you are lucky enough to have one) that fact may also go unnoticed.

And, of course, there are the usual farm attractions: space—you can be very "lost" when someone's looking for you to do dishes; beauty—which a child revels in long before he recognizes its existence; motion—always a tractor or truck or trailer moving in some direction; animals. Even work is not so painful here. It means joining fifteen other kids in a field and getting paid for labor that seems incidental to the general round of hilarity.

The haymow is what it always was to visiting children —a fascinating hideaway with unlimited possibilities. Stacking the hay bales to form a labyrinth of tunnels provides the ultimate in farm fun. (And leaves behind enough booby traps that the farmer does not forget you all winter long!)

No doubt about it, it is a fun farm for kids; and Uncle Paul, despite his 100-hour workweeks, recognizes the fact. So he takes them to market, treats them to breakfast, puts up with their noise and their endless questions, wipes their noses, ties their shoes, and enjoys along with Aunt Pat the privilege of being friend as well as relative.

But justifiably he grumbles. . . . "Next year," says he, "we're going to declare a cousins' week, and they're all coming at once. And I'm going away!"

"Would you mind very much," I whisper behind my hand, "if I came along?"

Rear Perspective

IF YOU LOOK OVER ALL THE HOMEWORK PAPERS carefully, read between the lines of the terse notes on the report cards, make a conscientious study of the moods and temperament of a grade-school boy, you may have some vague notion of how he's doing in school. But if you would open your eyes to a thousand wonders, go for an hour and sit on a too-small chair in the back of his schoolroom.

You may have thought you remembered grade school—a handful of isolated occurrences stand out painfully clear in your mind. But now you are here for the first time in the fullness of your perception. You even find your long-ago self in this vulnerable group, perhaps in the person of your own child, more probably in a combination of several.

Study these open, eager, guileless children, as like small sponges they absorb their vast culture, drop by drop by endless drop. You will discover new appreciation for the human mind, for the systematic arrangement and presentation of knowledge. And you will know something of the joy and rewards of this teacher.

Reflect from this seat in the rear that an elementary-school teacher must for nearly six hours daily react or speak in patience and restraint to thirty unique and groping individuals; she must deal with annoyances ranging from runny noses to sibling rivalry. She is expected to cope with and solve problems despaired of by harried parents, all the while teaching clearly, painstakingly, repeatedly the minutiae of the English language, the deceptive simplicities of numbers, the dubious skill of writing. You will regard with new respect this unsung heroine; her face will ever afterward haunt you in voting booths and gossip sessions.

But all this is "overlearning." The deep and lasting importance of this exercise in observation is what you will learn about your child and yourself. Watch the light of joy flicker over his self-conscious face when you come into the room, and you will know as never before the awesome responsibility of parenthood. Note the way he wears his pride on his sleeve as he carries a chair to his reading circle, and you will know more of the love he bears you than you may ever know again. If in the course of that hour he comes close enough to clutch your hand in a clandestine embrace, you will sense with a panic in your heart the depths of his need.

But begone—back to your pots and pans before the closing bell shatters the atmosphere of educational incubation. You may be sure that the independent little monster who comes slam-banging in the back door at 3:45 is no more kin to the innocent, open, loving child of 10 A.M. than is the fishwife bellowing, "For crying out loud, hang up your coat!" related to the sympathetic model of parenthood who sat this morning in the too-small seat at the back of the room and brushed tears from her eyes.

Neighborladies

A NEIGHBORLADY IN HER KITCHEN IS QUITE THE most fascinating subject for observation when you are a very small girl. As a child I often wheedled entry into the kitchens of my neighborhood. If there were children, of course, it was not necessary to stand on ceremony, but the kitchen memories that are the clearest are of homes where there were no children and one invited oneself. I would knock and then stand dumbly until the neighborlady let me in, whereupon I would sit as dumbly watching.

The Kalbach kitchen had the strangest fascination of all, for it was no place like home. It was nearly devoid of furnishings and scrupulously clean. There were only three straight chairs and a bird cage in that big white room with the scrubbed board floor. I had the constant sensation of having come on cleaning day, only later realizing that in this old German household every day was cleaning day.

The Kalbachs spoke little English, so communication with them consisted only of friendly self-conscious grins. Always they offered me milk, fresh and warm from the cow, which I firmly declined. Compared with our noisy, crowded, disorderly household, this kitchen offered a strange peace, which I enjoyed in small doses.

Mrs. Babcock's house down the road and across the creek was a warren by contrast. She was an unkempt eccentric who kept a mangy old Irish setter named Jip. Her tiny house, which smelled strongly of dog and neglect, was crammed with dusty furniture and bric-a-brac. I was somewhat frightened by her and the dog, but there was an evil curiosity about the place so I went there anyway. Rumor had it that she never washed her dishes, but only ate on the dirty plates of the previous meal. She was ill at ease during my visits and kept up a near monologue of prying questions, never failing to ask, "You got your cow yet?"

I think that Mrs. Diedrick next door felt overrun by the little Penton kids. Her kitchen was nearly as clean as the Kalbach's, but much more comfortably furnished. I liked the look of her shiny gray gas stove, which was so much nicer than our own kerosene variety. The whole house smelled, however, of escaping gas and vinegar. One never sat long in that kitchen before Mrs. Diedrick would say, "I think your Ma wants you. I can hear your dinner bell."

I think of those long-ago visits when my little neighbor girls come by for an afternoon of discovery. Karen and Leslie gain polite entrance by bringing in a pan I have set out for the dog. They know by intuition that a polite "Thank you" means "Run along, no time for little girls today." But they hesitate hopefully for some remark that translates, "Come in and visit." Or if my mood indicates that maybe after awhile I'll have time, they will sit on my steps and wait with incredible patience.

Their open, eager faces make them nearly irresistible to the mother of brash and uncooperative boys. Willing little hands and feet vie for the "privilege" of carrying out garbage or running errands upstairs and down. I take shameful advantage of them and they are delighted. But if I find nothing for them to do, they are content as I once was to sit and watch my every move as spectators watching the ball in a tennis match.

And finally they will ask shyly if they can play with the doll dishes. A yellowed box of doll dishes once stored hopefully by the mother of sons, now only stored, is a rare treasure to two little girls.

"Can we take them home, Aunt Pat?" they ask, and I am tempted to consent. That would, of course, destroy the magic. At home they would be mere toys to tire of and neglect. In the home of the neighborlady they are objects of wonder. The dishes I would willingly forego, but the joy of seeing my two little neighbor girls absorbed in their make-believe, serving a meal on the piano bench, *that* I cherish.

"No, I need them. Put them back on the shelf and come over another afternoon. I think I hear your mom calling you. . . ."

Ghost Story

THE FIGURE SILHOUETTED AGAINST THE FULL moon, stretching its arms in a swath of white over the sleeping child, is neither ghost nor witch. Rather, she is the good fairy of All Hallow's Eve, who sits up red-eyed and weary, fashioning a "friendly ghost" out of a worn feed bag sheet and last year's clown hat. Now she steals into the moonlit bedroom to measure the garment against the outstretched form of her child.

Years of this folderol had nearly convinced me that there were simpler ways of coping with costumes. When Teddy, hanging his head over the front seat of the car one day, peevishly asked when we were going to buy his costume, I was ready to head for the nearest discount store. Dane, however, spared me the trouble of answering, and at the same time redeemed ten years of midnight labor.

"We don't buy costumes in this family, Ted. Those things from the store are cheesey. Mom can make you a better costume than any you can buy." It had the ring of big-brotherly authority, and I was not about to spoil things by revealing that for as many years as Dane had packed his Halloween costume off to school in a paper sack, so had he coveted a boughten costume. Back to the sewing machine, the scrap drawer, and the labor of love.

The first decisions on matters of masquerade are made in the rear of the school bus, where in simultaneous monologues the kids announce their big plans. There is more talk as they hang up witches and paint orange pumpkins in home room. The final costume theme matters little; it would always have been better to have been something else. The really clever thing to do, I have often thought, would be to find out who sits next to your child, call his mother and discover what he's wearing, and then proceed to dress yours likewise. This would please him more than if you spent three years forging him a suit of armor.

Orrin wants to be a ghost, which is pedestrian enough. The white shapeless sheet that has successively robed an angel, a fairy, and a shepherd will adapt nicely to ghost.

"I'm gonna' be a skeleton," announces Ted as I imagine myself struggling with rib cage and sacroiliac painted on black-dyed underwear.

"How about a nice Peter Pan costume?" I ask, concentrating on the reserve of things in boxes upstairs. "Or maybe a devil?"

"Peter Pan!" he exclaims as though I had suggested Bo Peep.

"Well, would you believe Robin Hood?" I continued.
"Naw, I wanta be a skeleton."
"Why don't you just try it on and see how you look?"
Only because it was bedtime and would permit an acceptable stall did trying on a costume appeal.
"Hey!" cried Orrin, coming upon the green-clad yeoman, "Look at Ted! He's gonna' be the Jolly Green Giant!" From then on it was just a matter of finding a can of peas.
But just in case he changes his mind in the morning, I've laid out the devil—and as Daniel Webster could tell you, that's a pretty good trick even for a Halloween fairy.

Reflections of a Tooth Fairy

"CAN ANYTHING LIVE ON ANOTHER PLANET?" was Orrin's first question this morning. Later he sat on the floor with a puzzle of the United States and asked if Indiana is east or west of Ohio. It was while he worked his equations after supper that the tooth came out.

When eleven minus six equals ten minus six plus one, it might seem illogical that a tooth under a pillow plus a fairy equals a quarter, but Orrin has faith. Despite all the scientific method and moon exploration have done for his field of knowledge, he manages to keep a foot in both worlds, the real and the make-believe.

A child doesn't grow up believing in the tooth fairy as he does in Santa Claus and the Easter Bunny. She imposes herself upon him when faith in the other two is either very shaky or has collapsed altogether. Whether or not he believes in her is a test of his capacity for faith and her performance as a fairy. (When the tooth's been under the pillow for three nights straight and nothing has happened, any child will turn skeptical.)

But just as surely as eleven minus six equals ten minus six plus one, the scientific method and the fact of life on other planets will eventually play a part in shaping Orrin's faith, not only in tooth fairies but in God and man and their coexistence.

The "fairy" who steals into his room, reaches carefully under his pillow for the first baby tooth, and leaves a shining coin is aware that this child must believe in more than a silver coin if he is to survive in the struggle of faith. He must believe in a set of ideals, and he must believe in other people. Most of all he must believe in himself; and that's a responsibility that falls directly upon all the "tooth fairies" in the world.

It's not essential, of course, that children believe in tooth fairies, but it's terribly important that "tooth fairies" believe in children.

What's With All This Santa Claus Jazz?

THERE IS DEEP PATHOS IN WATCHING A MYTH die, and anyone who has stood with a six-year-old beside a Santa Claus not too professionally done will feel the tug of the experience on his heart. When the child sidles up to you and whispers the discovery that the tummy is really a pillow, you know he clutches frantically to the hope that he is wrong. He will dwell hereafter on other aspects of the legend. Perhaps this one helper is a phony, but surely the reindeer and all that stuff . . . ? Well, we'll leave out some hay Christmas Eve and find out.

Experiencing it all, you are suddenly aware that there is much more at stake here than the destruction of a legend. Here is a child who may soon see himself as the victim of multiple conspiracies. You may do what you will to extend his faith, but there is a realism that stalks the schoolyard. Sooner or later he'll come home and tell you, "Ronnie says there isn't any Santa Claus! We told him there is! We're right, aren't we?"

The dilemma of the Christian parent is to inject into the Santa Claus legend an element of the finite and the infinite, to establish a correlation between a pagan myth and a Christian truth. It's best to be prepared with answers, for you never know when you'll be confronted with a crucial conversation. This one came as do so many of our crucial conversations, from two little boys hanging over the front seat of the car.

"How does Santa know what I want for Christmas? I wanted roller skates last year and he knew. How does he know?" asks Teddy with candor and no real doubts.

"Well, Santa Claus is a spirit, and a spirit can be all around us," I answer. It takes a little time for this to soak in.

"Mom, what's a spirit?" he asks.

Theologies have risen and fallen on this question, and a parent can only stammer over it while he prays for someone to change the subject.

But Orrin is younger, his vision closer to the infinite. "I'll tell you what a spirit is, Ted. It's a 'knower,' isn't it, Mom?"

"Yes, dear," says Mom, dispatching a small prayer to the Holy Spirit. "It certainly is a 'knower.' "

Teddy, faith unshaken, now has the piece he needs to solve the puzzle. "And then he turns into a person?"

"Yes, dear. Once a year, at Christmas, he turns into a person." The Word became flesh.

Snowshine

Many a cold, miserable night bound on a dutiful mission somewhere I have gazed through lighted windows (like The Little Match Girl) at people sitting tranquilly in warm living rooms and wished I were home doing likewise. Tonight was not one of them.

Orrin and I, dressed warmly in ski clothes and sheepskin boots, took our sled and walked a winter mile down the road to go coasting. We looked through neighbors' windows and regretted that they didn't share our fun.

The road was slippery and powdered with freshly falling snow. We hung onto each other for balance and struck up a marching cadence: "Left! Left! I had-a-good-home-but-I-left! Left! I had-a-good-home-but-I-left! Left! . . ."

"Did you?" asked Orrin.

"Did I what?"

"Did you have a good home?"

"Sure I had a good home. What do you think makes a good home?" I asked.

"A good home is where a baby can be born and grow up and everyone loves it," he said. And he wasn't talking about just any baby; he was talking about Orrin. I hugged him a little closer and we walked on.

The coasting with the neighborhood children and some of their hardier parents was exhilarating. Russell Schmalz brought out his tractor to throw irregular light on the slope. Then for two hours we tumbled downhill on sleds or "flying saucers," or flew blindly through the light snow on Christmas toboggans. The children discovered that they could even slide down on their feet or upside-down and backward on their slick nylon parkas.

At ten o'clock Orrin and I said goodnight and made our way back up Bank Road. We took turns pulling one another on the sled over the fast new snow, and when we came to a downgrade we both hopped on and coasted.

For most of the way the road parallels the riverbank and is separated from it only by a wide fringe of trees, bare and black but etched against this night by white pinstripes of snow. At one point we left the sled, waded through the drift at the road's edge, and trudged to the valley's rim to look down to the dark river below.

"If you're very quiet," said Orrin. "you can hear the water." So we listened intently to catch the faint roar.

"Listen to this echo," he said, and he whistled a funny sort of whistle that a boy must be very proud of when he's missing two front teeth. But I didn't hear the echo. . . .

"Do you know what would make this picture perfect?" he said, indicating the quiet bit of woods around us. "A mother deer and her fawn."

I was the "deer" and Orrin was the "fawn." As far as I was concerned, it was perfect.

At the End O' Way our old farmhouse was silhouetted

against the sky by the light of the back door. Peppy ambled out on her arthritic haunches to meet us at the mailbox and wag us home. Then a welcome hot bath and bed.

As I tucked Orrin into his top bunk, he shared with me a little of his nine-year-old philosophy. "Do you realize, Mama, that this day is gone and will never come again?"

How well I knew! This day was gone, this beautiful January day. It was a rare circumstance that had left the two of us alone tonight to go down the road coasting. Perhaps never again would just we two walk down that road on a winter night with arms around one another, snow tickling our faces, talking about a curious range of subjects—what makes a good home, why nuns pray so much, why some houses have icicles and some don't.

"Yes, Orrin," I said, "I do realize that this day is gone and will never come again." Finally I was hearing the "echo"—my heart echoing his.

And One to Grow On

IF A MOTHER STANDS IN THE TRADITION OF martyrs she will rise unflinching to the occasion of her son's first real birthday party. And that is the last time she will assume that pose. Birthday parties are recognized thereafter as occasions for cowering.

You may practice deafness during the weeks before birthdays, pretending you don't hear the party plans in process. But on the night before when your son hands you a list and says, "These are the guys that are coming to the party," you have to make a choice. Either you call the mothers and acknowledge the party, or you call and say, "We are *not* having a party tomorrow!" in which case you come off as a pretty ugly mother.

If you proceed as though there is no party and call no one, you may be in for a rude surprise. One of my friends reports the trauma of having ten little boys get off the school bus to come to a party she hadn't anticipated!

My husband takes a dim view of my confirming a party

by phone at 9 P.M. the night before. "Just where are those mothers going to find birthday presents this time of night?" he asks.

"That's their problem," I mutter. "I have worries of my own."

My first birthday party was planned to impress mothers, with a theme and decorations and favors and food à la *Ladies' Home Journal.* Five minutes of that fiasco convinced me that the time should be invested in planning games, and the money in prizes. (Especially when it's winter and you're confined with fifteen bloodthirsty pirates with wooden swords.)

If you plan twice as many games as you think you can cram into two hours, you may succeed in filling fifteen minutes of the first hour. I have spent hours waiting outside operating rooms, days in dentists' chairs and labor rooms, but I have never known time to pass so slowly as it does during the games at a boy's birthday party. Everything is in motion but the hands on the clock.

A scavenger hunt calculated to consume time is over before you've finished scattering peanuts for the peanut hunt. Boys who take twenty minutes to bring a jar of jelly from the cellar can in five minutes locate three eagle feathers, a rusty horseshoe, and last year's truck license, if there's a contest involved.

Try to stretch a game five minutes longer than enthusiasm dictates, and while you're frantically icing a cake in the kitchen, some honest child will stab you with, "This is no fun. Let's play what we played at Walter's party."

If there isn't a prize for everybody—especially your own child, who is the only one acting his ugly self—the party is a crashing failure. It must be made to seem that there are no losers at a birthday party. Although when the last balloon is broken, the last boy has gone, and you stand amid the debris—ice cream melting on paper plates, cake crumbs on the rug, peanut shells and torn paper everywhere—it is obvious that there is always *one* loser!

Anybody for pin the tail on the donkey?

Middle Size

Our younger boys are now nine and eleven going on sixteen and eighteen. Our house seems joyously overrun with "neither nors" treading a happy path between childhood and adolescence, demanding as is convenient the rights of both, accepting graciously the responsibilities of neither. I observe that this age group, roughly nine through twelve, composes a more or less specific classification of the genus Boy, and elicits a number of rather well-defined characteristics. For want of a better term, I shall call this classification Middle Size.

Members of this group, for example, feel entitled to wear their hair down to their eyeballs, but feel no obligation to comb it. They insist on "cool" clothes, which they will fish from under the bed for the third or fourth day's wearing. Middle Size boys operate on the basis of one inflexible rule: Get everything out, put nothing away.

Middle Size friends call on the phone and in a determinable number of minutes (the time required to pedal from there to here) will enter the house, make their way to the small bedroom upstairs where we seclude our TV set, and caucus. Then they spill over to all parts of the house in pursuit of varied activity. If they launch into a peaceful game of Monopoly, it subsequently erupts in a noisy hassle between the brothers, splitting the group into factions bent on outinsulting one another. Ultimately they will straggle to the kitchen where they display a characteristic common to all species of this genus: insatiable appetite.

Whereas small boys can be satisfied between meals with a few cookies and a glass of milk, Middle Size have appetites leaning more toward large pots of macaroni or spaghetti, one pan per boy. When they become aware that their friends are sitting around salivating, they inevitably repeat the cooking process for each friend, leaving the kitchen a wasteland of starch and strainers and gummy pots.

Dare to suggest that they probably won't want much supper and you are treated to a harangue on how bad the cafeteria lunch was today. "All we had to eat was chocolate pudding!"

Bellies full, activity exhausted, they adjourn to a neighboring house where I presume they overrun the premises in like fashion.

By the age of nine or ten a boy has deduced that honesty isn't necessarily the best policy, yet he feels a certain commitment to truth. Middle Size, therefore, becomes the age of the half-truth. A statement like, "David says his grandfather's going to buy him a car," could be accepted as an indication that the old boy is out of his mind. It may mean that for his birthday David is receiving a 6″ × 2″ model car. It probably means that Grandpa was reading his newspaper when his grandson made the preposterous request, didn't hear the question, and made the mistake of grunting, "Umphh."

Middle Size is an age of persecution; always in need of interpretation is a statement such as, "My bus driver hates me." It can mean that she is a nasty and vindictive woman who is persecuting your son. (An explanation too readily accepted by parents who feel persecuted themselves.) It could mean that she had a bad day and looked at him cross-eyed. It probably means that after warning him eight days arunning, she made him sit in the front seat for sticking his arm out the window.

Girls with thin little voices begin to telephone at Middle Size. And boys with sheepish grins answer in their high-pitched tones, "Yeah, whaddayuh want?" Middle Size boys in general scorn girls, while their brave hearts beat wildly in timid breasts and they covertly inscribe initials on their forearms.

Love is best expressed during this period in harsh phrases like, "Boy! I can hardly stand Mary Sue Thompson!" But a boy's friends will freely give you the straight scoop while he beats them to silence and tries to conceal a smile. Having a bevy of a boy's friends around always helps in cutting through the half-truths. More truth emerges around a noisy supper table where Middle Size boys are vying to top one another's facts than are ever told in group therapy sessions.

Middle Size has a garbled body of sex knowledge always gleaned from innocent kids down the street. Middle Size friendships run hot and cold. This week's friend is next week's arch enemy. Middle Size boys never have homework until 9:30 at night, and getting them to bed is like trying to settle a war in the General Assembly of the United Nations.

Middle Size boys still want you to share their prayers, but they don't want the fact known at large. They have abandoned Santa and the Easter Bunny, but are devoted to perpetuating the myths for younger children. Middle Size children scorn the responsibility of younger siblings in the presence of their parents, but boisterously assume authority when their parents are out of sight.

What boys want most at Middle Size is spending money and minibikes. What they need most is understanding, responsibility, and plenty of attention. And you may be sure that if their needs aren't met at home, they will go to obnoxious lengths to win attention elsewhere.

A *Toast to Mothers*

LONG BEFORE MY MOTHER'S FACE CAME sharply into focus in my mind, I could have described her hands. They were small for one who seemed so large—the skin stretched tautly over her knuckles. Over the backs of them where the blue veins swelled and relaxed with the motion of her fingers, the skin was shiny and covered with masses of freckles.

My mother's hands were more at my eye level than her face, but surely my awareness of them bespoke the loving care they extended. I remember those hands washing my smaller ones as I sat in the kitchen sink with my feet in a wash pan. I remember their cradling me in an armchair as she read to my brothers and me. And I recall their touch on my warm forehead as I lay ill on the couch. I remember her hands at the table mashing my potatoes, pouring the tea, working at the washboard and the wringer and the clothesline. I remember those hands all flour and dough as she kneaded the bread, scattered feed for the chickens, pulled sweet corn in the field. My mother's hands gave me the secure feeling that the world existed for me and that I would never want for anything.

No mother need ever belabor the point that she works hard for her family. A small child defines his mother in terms of the many things she does for him. One little girl in my son's class groping for the scope of her mother's ministrations says, "She does so much I can't even say them all."

As I grew older my focus on my mother shifted from her hands to her face. I knew by then that it was a big world, that I did not have this woman all to myself. But it didn't seem to matter. Her response to each of us was wholehearted—one child, one mother. It is in this acknowledgement of a child's "gifts" of gratitude, a handful of short-stemmed flowers, a clumsy attempt to make a bed, a wrinkled arithmetic paper with a star, that a child finds or fails to find the acceptance he longs for.

It was a well-guarded secret between my mother and me that I was princess of the earth. When I discovered—very soon—that everyone didn't share that view, I did the best I could to conceal the fact from Mother. I could bear the truth that I was just another girl among girls, but I wasn't sure that she could.

The truth was, of course, that my mother had two princesses and five princes. No one has ever convinced her that they were otherwise.

Love certainly works her greatest miracle through good mothers—their serving hands, their gentle approving faces, their unswerving faith in their children. There is no gift a mother gives a child more important than believing in him so strongly that eventually he comes to believe in himself.

4 THE PAST

The lines have fallen for me in pleasant places; yea, I have a goodly heritage.
PSALM 16:6

When you are the fourth generation on the "home place," the past stalks you. It lingers in the barn lofts and blows beneath the clapboard on icy winter nights. It lives indelibly in the minds of the relatives and the old residents. Unquestionably it lends a rich influence to our lives at End O' Way. Times and life styles change but so many elements are constant—the land, the buildings, the farming cycle. To these I bring my similar rural heritage, the influence of those secure years of my own past at Penton Orchards.

Anybody Else Get Confused?

A FEATURE OF FARM LIFE THAT INTERESTS AND amuses me is the system of geographical nomenclature that evolves in a rural neighborhood with the passing generations. When a farmer has half a dozen fields, designations get very complicated.

Frequently the place names bespeak something of the history of a farm or region. On our property there is a promontory with steep shale banks jutting into the river valley. Because of its natural fortification, it served for hundreds of years as an Indian encampment. To us it remains "the Indian Fort," and quite logically names the adjacent field.

The "woodlot" ceased to be a woodlot forty years ago. (We're growing wheat in the woodlot this year.) The "orchard" hasn't been an orchard for twenty-five years. Moreover, there's a field called the "old orchard" that hasn't been an orchard for seventy-five years!

That's complicated enough. But when a farmer acquires his neighbor's land, he transfers the terms that identified that man's property. Hence I must discern where "Newberry's woodlot" and "Newberry's orchard" once were.

Pastures frequently conform to the duller differentiations of east pasture, west pasture, back pasture, side pasture. Sarah Born, one of my dairy friends, tells me that before cows were computerized, there was always a "morning pasture" where the critters lined up at dawn for the milking.

"And doesn't everyone have a creek lot?" she asked. "One of our fields we called the slashing," she added.

"Why was that?"

"Gosh, I don't know. Just always was that."

Random House Dictionary identifies a slash as "a tract of wet or swampy ground overgrown with bushes or trees," and then in brackets at the end of the definition adds a question mark. Evidently even the dictionary staff aren't sure why that field was called "the slashing." Somebody stopping around for a bag of potatoes tells me that the term developed when the land was cleared through the practice of slashing all the trees in a piece of woods on one side and then felling them en masse.

Sarah hastened to add that the Ohio Turnpike now slashes through "the Slashing," thereby reinforcing whatever definition might be found.

But the largest number of fields or farms are classified under the name of the former owner. A piece may be "Aunt Annie's field," though you've paid the taxes on it for thirty falls and springs. It seems somehow fitting that the piece of ground where a man invests so much of his life should long after bear his name.

Consider the case of the Schmalzes who live on the "Bacon place," grow fruit on the "home place," corn on the "Blossom place," and wheat, oats, soybeans, and what-have-you on the "Kropf place," the "Kneisel place," the "Zunt place," and the "Brown place."

The Clines farm the "Morse place," the Grosses live on the "Baldwin place," and the Frys are buying the "Geggenheimer place," which is farmed by the Schmalzes.

No wonder I often feel like a stranger in these parts after only twenty years' residence!

The Good Old Days

A CENTURY AGO NORTHERN OHIO WAS AN agrarian Eden of broad tree-lined avenues behind which sat large, well-kept homes neatly and simply landscaped with well-trimmed yards surrounded by white picket fences. Horses and wagons skimmed along these wide roads, rumbled pleasantly over the bridges en route to mills where waterwheels turned with gentle splashing. The farmer at the reins whistled happily, thinking of the dear little helpmate humming over her shiny black range back in the spacious kitchen . . . or so one might think from the artist's illustrations in the *History of Loraine County* published in 1879.

It is the artist's prerogative to be selective in detail. And if you're trying hard to sell history books, better to establish a romantic point of view. Better to leave out the ruts in the highways that were rivers of mud in late winter, better to set the privies out of sight, catch the picture in early spring before the weeds started, get those geese, ducks, and chickens out of the backyard. What the farmer is thinking and what his wife is saying are only suggested, of course, by the tidiness of the whole.

There is also a myth that lingers about the milk of human kindness that flowed in such abundance "in the good old days." I've been delving lately into the facts of the matter and have grave misgivings.

Take the case of Mary Watts, whose reputation for being loose about the truth was dragged before the Board of Deacons of the Brownhelm Congregational Church during the winter of 1857. It seems that Widow Watts concealed from the administrator of her husband's estate the fact that she possessed some valuable silverware, claimed also that the clocks and knives in the household had been purchased with her own money. There were six other grievances against her, each a little pettier than the previous. The church elders voted to suspend her from membership despite her admission that she was "liable to do wrong and doubtless had," followed by her plea for forgiveness.

A year or so following the offenses, she appealed and confessed again, saying, "If there are any of the members who feel they cannot forgive me and are without sin or fault, let such first cast a stone at me." It was evidently a poor text she chose, for it was voted that "this confession is not such as the case demands," and she was never mentioned again in the records.

It is not difficult to imagine the years of bitterness and personal anguish that must have followed every such judgment (and there were many in that record book). Maybe Mary Watts didn't purchase the clocks and knives and other silver with her own money, but I'm sure she paid plenty in blood, sweat, and tears. I say the persecutors all deserved a scarlet "M" for "mercilessness." (Although John Locke, who seduced a young woman, was forgiven and reunited to the church within a year.)

I don't know if Widow Watts lived behind one of those neat facades in the history book, but I am not taken in by them. I know they often concealed the poverty of broken spirits.

"I Remember, I Remember . . ."

If you live on the "home place," it is to be expected that many people have a prior claim-of-the-heart to the farm you otherwise think of as yours. Many an aunt and uncle and cousin feels a stubborn reluctance to knock at the door that was always open to him; and, being in sympathy, you do not expect that he should. Especially during the reunion season, you feel a willing obligation to whip the place into shape, to trim the shrubs and mow the orchard, to haul away the junk and rake the twigs under the maples, to pull the weeds in the flower beds and sweep the sidewalks.

And then the relatives come. Perhaps it's for an all-out reunion, with everyone spreading cloths at tables under the trees (they'll talk of how they remember those trees as saplings). Or perhaps it's just an evening visit. Always fearful that you will think they intrude where it really seems their birthright to intrude, these affairs are arranged with utmost consideration for your convenience.

The uncles will be nostalgic about the land, about the spring and the old woodlot, about the logging trail and where the orchard stood, about how the crops have changed from what they once were. The cousins remember the barns (especially the haymow) and the river, the ice house, the woodshed, and the attic. They talk of summer vacations and the frolics they had together, of the deep winters and the sleigh rides.

The new generation rove about impressed only by the rather ordinary character of what they had been led to believe was some sort of Eden. When an overzealous parent with a sense of tradition tries to explain the significance of this or that landmark in his own life, the kids are busy turning him off.

But the women concern themselves especially with the house, mentally rearranging with a nostalgia that is almost sadness. When an aunt pokes her head through an inside doorway and gazes slowly about the walls and the ceiling, you know that she does not notice the cobwebs you've overlooked. Nor does her gaze rest on the wallpaper or the Danish plates you prize. What she sees is how things used to be in the sitting room—the stove with its black chimney (no longer so ugly in her mind's eyes), the black leather sofa, the plum-colored reclining couch in the corner under the gas light.

"Over here," she'll tell you, "was Grandpa and Grandma Leimbach's bedroom. There was a window there where your kitchen door is."

"This was the parlor," she sighs, retreating to the living room and sinking wistfully and somewhat suitably into your rattan "womb" chair. You realize how profane it must seem that the carpet is worn to the warp in here and the children's toys are strewn about. "We only opened it for weddings or funerals or the preacher's calls," she says.

We like to think that we have improved the place, but we are wise enough to recognize the relativity of an "improvement." Is there anything *Better Homes and Gardens* can offer that will improve the "homestead"? When Cape Cod curtains replace the starched lace that hung at the windows, when two acres of lawn flow where once a picket fence enclosed a world, when geraniums bloom in place of hollyhocks against white clapboard, when trimmed evergreens take the place of sprawling lilacs and syringa, could anyone call it an improvement—if Grandma and Grandpa are gone from the porch?

"P" Is for Peabody

MANY A CHILD HAS TAKEN HIS FIRST POKE AT A bully defending the anonymity of his middle name. Parents who celebrate their heritage in unusual names are relying on a conviction that by the time the name becomes an item of public issue at the commencement exercises, their child will have grown up to supporting it.

We "suffered" such names, my brothers and I: Erik Musselman, Theodore Purvis, Henry Simpson, William Dean, Patricia Ryall—certainly not the kind of name you want whispered around the fourth-grade room, but the sort that great aunts delight in and sometimes remember to inscribe in their wills. I hasten to say that all I ever got out of Ryall was four tablespoons with a script "R" and a silver thimble that fits my thumb. Brother John, who, thanks to Mom's warm memories of an old beau, was called John Alfred, judges he may have lost a fortune for the simplicity of his name. On his first visit to a wealthy maiden aunt in Los Angeles during the war years, he was obliged to argue at great length to convince her that the "A" was for Alfred. Surely, she insisted, they wouldn't have named him Alfred when the family tree hung so heavily with John Augustus!

My sister Mary Alice was also very simply named—something to do with a whimsy about Alice in Wonderland. But as for the rest, our names gave us a sense of having derived from somewhere, of having a responsibility to a tradition. And so it is that though you have repeatedly raised a fuss about the name or kept painfully silent to spare your mother's feelings, when you walk across a commencement stage it seems at last quite right that you should be called John Arbuthnot Smith. It is not so much a burden to be borne as a tribute to fulfill.

What Is a Big Sister?

A BIG SISTER IS AN UNFORTUNATE SOUL SENT into the world to smooth a path for you.

A big sister is a buddy until she outgrows dolls and is installed as your baby-sitter.

Big sisters are the little sisters their mothers never had.

Big sisters wear their first lipstick, their first heels, and have their first dates two years later than the sisters who follow them.

Big sisters have to be *in* at about the same hour that their little sisters will be going out.

A big sister is a sex-education teacher who whispers the facts of life across the gulf of darkness separating twin beds.

Big sisters write you long letters full of confidential information.

Big sisters remember your birthday in wonderful ways.

They will walk through fire for you, but would rather die than admit it.

Nobody ever does as much for big sisters as they do for little sisters.

Big sisters love babies, big brothers, big brothers' friends, convertibles, and typing.

Big sisters hate little brothers, doing dishes, baby-sitting and algebra.

You can borrow anything from a big sister except her boyfriend.

A big sister grows up and marries a brother-in-law with whom you are secretly in love.

Big sisters make the best housewives, having served such extended apprenticeships.

Big sisters always solve problems.

Big sisters never understand why their own children are neither as industrious, conscientious, frugal, nor maligned as they were.

Big sisters are people who take charge in emergencies.

No one understands each member of the family as well as a big sister.

No one can scold you with such effectiveness, nor forgive you as quickly as a big sister—nor continues to do so for as many years.

A big sister is a queen who cannot abdicate.

It Takes a Heap o' Living . . .

SOONER OR LATER EVERY CHILD COMES TO HIS mother complaining, "There's nothing to do around here." This is followed with a plaintive, "What can I do?" After a mother has made half a dozen suggestions that are met with a whine, she's usually tempted to suggest that the kid go out and play in the traffic. Whenever this happens to me, I think with nostalgia and a certain burden of guilt of the play activity in my childhood. Sadly, though, there is little of it that I really want to suggest to my children.

My mother kept a different sort of house. There was no family room, no recreation room, no play room. All seven of our rooms were "living" rooms, and the only priceless objects in them were the children.

In the kitchen we played darts with the back door as a target. Because the floor slanted drastically from north to south, it was a good roller rink for beginners. My personal amusements there, to the annoyance of everyone else, were hanging on the roller towel and swinging on the dining room door.

Our dining room was the center of the house. Everything led into it and nothing, it seemed, passed through. The table was usually piled high with clutter, but sometimes we cleared it, put in the extra leaves, and played Ping-Pong. Small matter that it was a round table.

The scarred and battered baseboard in the corner by the bathroom door was the backstop for the marble ring. The marbles glancing off the walls here clattered toward the front door and were lost in the cold-air return. The dining room linoleum was periodically trimmed to fit the kitchen and replaced by a new one. No one who has ever done it will forget the thrill of belly-slamming across new linoleum on a sofa pillow.

We lived in an era of sofa pillows and, ah, the great fights! For all her indulgence, Mother did frown on pillow fights as well as bed jumping. The boys' room had five beds arranged in such a way that we could almost circle the room on them. We never split a pillow, but we went through a lot of bed slats.

Hide-and-seek was a popular indoor game with us. You could seldom be found if you wedged yourself behind the chimney in the closet under the stairway, but the best place to hide was on the top shelf of the bathroom closet with the bedpan.

We often constructed tunnels by turning over all the living room and dining room chairs, or draped the library table and built a cave. Visiting children were always fascinated with our Morris chair, and we were never reluctant to give demonstrations, stretching it out to full length, pushing the arm button, and catapulting forward. In more subdued moments we built houses of cards, or castles out of dominoes. We also wore out a couple volumes of reference works using them for building blocks.

My mother only came to the study of psychology in recent years, but it must have been gratifying to her to realize that she had in no way frustrated in their play the "tender blossoms" that grew underfoot. She seldom scolded and never nagged. She ran a wreck of a house, but it was a great home.

Once Upon a February 14

SOMETHING STRANGE HAS HAPPENED TO VALENTINE'S Day. It used to be celebrated on the 14th of February, but now it's on the Friday the teachers deem most convenient to that date.

In my day there was great rivalry over the number of valentines the kids received. As a chosen few distributed the cards from that magnificent box up front, you sat and watched your stack accumulate. When it seemed that a girl across the aisle was doing better than you, your heart sank. Then perhaps there would be one, two, three for you —and you were catching up. As surely as you came home with forty-three valentines, your sister had forty-four, or she knew of some fourth-grader who got twice that many.

But all that has changed too. Teacher sends home a list, and in the name of fairness everyone's name is inscribed on a white envelope. (If you got a valentine in an envelope when I was a kid it was a big deal.) It's an adult way of handling the situation. Apparently there are some adults around who remember how it felt to get the fewest valentines in the class.

As I recall we couldn't always afford a valentine for everyone, so selectivity was essential. The messages on the cards were read with scrutiny. You had to be careful not to send anything too mushy to the "just nice kids" on your list, nor did you want to be too committal about your feelings to the person you'd been drooling over all year. The matter was usually settled on the basis of size; the bigger the card, the deeper the sentiment. You picked out the biggest, most noncommittal card you could find and let the fellow judge your feelings.

If, on the other hand, you received a card a foot high from some "creep" you couldn't stand, you folded it, or hid it, or did what you could do short of destroying it (not to reduce the total) to conceal the fact. Sometimes with a little calligraphy you could give the impression that an upperclassman with the same first name had a secret passion for you.

And oh, the smug pleasure of getting a valentine from somebody in another room! A kid you knew walked boldly in, dropped a single card in your box, threw you a knowledgeable glance, and strode rapidly out. Big thrill!

This was pre-room-mother era, but our teacher managed to give us a party. My good mother, bless her, always sent a box of cookies on Valentine's Day. That was something to make a kid burst with pride, bringing in unbidden a box of heart-shaped cookies decorated with red sugar.

Teacher printed a big VALENTINE'S DAY on the board, and we tried to make as many words as possible from the letters. Sometimes we sang, and we ate the candy she provided from her meager salary but the most wonderful thing about the final hour of our Valentine's Day was that Teacher let her hair down a little, acted almost human and approachable! On these rare occasions the braver kids would venture to ask her half-personal questions about her mythical existence beyond the classroom. Gee! Maybe that's it; maybe that's why Valentine's Day now falls on Friday. The teachers have learned that when you've come down from Mount Olympus, it takes at least a weekend to reestablish your deity!

Sense and Absence

One of the things we bragged about at our house was that Mama was never sick. If she had mornings when she would rather not have left her bed, she kept the fact to herself. We grew up with the conviction that you got up in the morning and went about doing what there was to do—which for us was chiefly going to school. Among the seven of us we logged more than sixty-five years of perfect attendance.

It was considered altogether feasible that you could absorb the three R's with a toothache or a boil or an infected hangnail. We hobbled to school on crutches or with an arm in a sling, with a bandage around the head, or a pain in the lower back. Malingering was unthinkable. Ted and John had on their records several cases of P.M. truancy in very late April or early May, but even these were considered more forgivable than malingering.

It was not all grit, certainly; we were healthier than the average family, and that too was Mother's doing. She schooled herself in nutrition and pursued it with a vengeance. Long before the era of vitamin pills, we were daily plied with cod liver oil; we drank hundreds of quarts of tomato juice yearly, and gallons and gallons of apple juice. We consumed a peck of apples and a peck of potatoes every day. Nearly all our bread was wheat bread, and wheat germ was smuggled into everything in which Mama could disguise it.

We lived in a drafty barn of a house without benefit of thermostat and slept in frigid, well-aired bedrooms. On weekends and after school we were herded out of the house for work or play, whichever exercise our age prescribed.

There were of course the measles, mumps, chicken pox, and now and then a case of whooping cough to keep one home from school. But if you were fourth or fifth in line, you stood a pretty good chance of catching all these things before you even started. This was Mary's good fortune, and she never did miss a day of school.

There was one memorable incident when Mom locked horns with a local doctor and ultimately the public health officials. Somebody noticed Bill had a rash on his neck and sent him home, though he felt perfectly well. The doctor said, "Roseola," and Mom said, "Nonsense!" She took him to another doctor who agreed with her, and Bill went back to school. The county health nurse called to investigate the situation, but Bill stayed in school.

I got the distinct impression that roseola was not allowed in our house, nor were headaches or female complaints. The common cold had to be pretty uncommon to warrant bed rest. Nothing short of a fever was an excuse for staying away from what we early came to regard as our serious responsibility.

There was probably a great deal of public comment upon the fact that we were all in school on the day after Daddy died, being excused only on the afternoon of his funeral. "Irreverence," they may have called it. Mother, who was never much concerned with what people were saying, would have called it, rather, "reverence for duty." It was a rule of life on which they had never differed, she and Daddy.

Attendance awards by today's standards are considered square. There's a great deal said about the epidemics spread by kids who would be better off home in bed. I seriously doubt the truth of this. The epidemics seem to flourish even with the policy of permissive absence more generally practiced by parents and students today. And this policy of permissive absence spreads throughout our nation a "disease" more insidious than any that normally flourishes in an overheated classroom, a disease of dishonesty and apathy. It's called "chronic absenteeism," and it merits everyone's serious attention.

Sassafras Season

"WHAT COULD I DO TODAY?" ASKS ORRIN, GAZing idly out the window on the first day of spring vacation.

"It impresses me as the kind of day to dig sassafras root," says I.

"Whats sassafras look like?" he asks.

Gosh, what does sassafras look like? It doesn't "look"; it smells and tastes. But as for looking—well, you just have to go hunting with someone who knows it. And suddenly I know what Orrin and Ted and I are going to do this afternoon.

Johnny and Bill always identified the sassafras. I just tagged along to carry the shovel. We pulled on our galoshes, for this was the mud season, clipped on our earmuffs, and plodded off in the chilly afternoon. We walked up the lane past the asparagus bed, past the peach orchard, through the rows of red astrachan trees where next summer we would climb for the first apples; then back through the baldwins, past the blueberry bushes and the rock pile. There were two rows of golden delicious trees west of the rock pile, then came the fence row between Dietrich's orchard and ours where we dug for sassafras.

John dug, and Bill and I knelt to break away the dirt clods. Once the shovel cut into the fragrant root there was no mistaking it for the other sort of scrub brush that grows in neglected fence rows. Digging was difficult in the nearly frozen ground, and we seldom dug more than a couple bushes. Fingers grew cold quickly in this chilliest, dampest time of year.

We carried the dirty roots home and scrubbed them in the sink, spattering half the kitchen in our clumsy efforts. Then we chipped the reddish bark from the heavier ones, being careful not to include too much of the bitter sapwood. We put it in the special "sassyfras" pan that through the season carried a red stain from repeated brewings. Between tea sessions the bark dried in that pan on the back of the stove. One batch of bark was good for a week or so, but each brewing required longer to achieve the desired shade of red.

Surely the spicy flavor was appreciated the most in that first hot cupful on the day of the digging. Diluted with milk and liberally sweetened, it was a spring tonic indeed as dusk closed in and Mother bustled about the warm kitchen getting supper and cleaning up after her muddy little kids.

You can buy small packets of sassafras bark in the supermarket, and I have done so a time or two. But the effect isn't the same. And once in recent years when Johnny was trenching for a waterline he brought me, on a nostalgic whim, a whole car trunk full of sassafras root. Even that didn't produce the old results.

Sassafras tea isn't just a drink; it's a ritual that calls forth its pleasure from the camaraderie of small kids with cold hands and feet and a warm sense of accomplishment.

A View From the Dump

WAITING FOR MY HUSBAND TO LOAD THE planter at the end of the potato field, I frequently sit on a boulder and look out over the deep ravine that through numerous decades has fallen heir to the Leimbach junk. There's a strange satisfaction in knowing that you can provide a depository for your own debris. It belongs to the sense of independence that has always been a farmer's pride.

The view here is not unpleasant, backed as it is by the tops of the sycamore, the willow, and the walnut trees growing in the river bottom. As I look over the accumulation near the top, I recognize it as a friendly bunch of junk. An old wringer washer carries me back to my initial struggles with a family wash. A wash boiler farther down the hill calls to mind the fact that someone struggled earlier and harder than I. There are old farm implements down there too, a reminder that farming has changed even more than housekeeping—a cream separator, a horse cultivator, and a beat-up hand duster.

The tin cans and broken bottles have settled beneath bundles of rusty fencing and reams of baling wire. A dense growth of Virginia creeper does what nature can to obscure the wastes of man.

Perhaps the saddest items in this dump are the assortment of discarded toys, toys never really appreciated to the full, hence neglected and abused. More than anything else here, they are symbols of how a little affluence alters a way of life. It is the toys that transport me in thought to another dump. . . .

Down at the north end of the pasture where the creek funneled into the culvert and ran under the "back road" was an unhappy out-of-the-way spot where one bag of rubbish invited a second and a third and finally a steady stream. No matter with what anger and disgust the property owners regarded the wretched heap, to the kids in our neighborhood it was a source of wonder and delight.

The contributors to this eyesore were a faceless lot, malnourished, we judged, on canned milk and whiskey. As we pored over the fascinating pile, sorting out a host of marvels, we concluded on the basis of a frugal set of values that they were also a shamelessly wasteful bunch.

Back along the creek on a gentle slope within clothes-changing distance of the swimmin' hole was a clearing in the thorn apple trees that served us as a playhouse. Here we carried the junk, and in our innocent pleasure transformed it into the treasure of childhood. Cracked plates and bowls and rusty pie pans, soiled lamp shades, chairs with missing rungs, old bedsprings, granite pans, and leaky washtubs—all were valued furnishings.

There was only one good reason for wearing shoes in

those long-ago Depression summers. It was only with shoes that you could tramp condensed milk cans into "horse-shoes," so we carried them home where our shoes always were and where there were surfaces hard enough to make a good racket. Along with the milk cans we frequently took some special prize too precious even for the playhouse, and always defined to Mother as "perfectly good yet."

Mother may have taken a dim view of our playing at the dump. But if there were lectures on the evils of germs and rusty nails, they were not forceful enough to dissuade us or to stay in the mind through the ensuing years. It was perhaps avant-garde "creative play" accomplished very simply without the hoopla of free schools and the open classroom.

La Vie en Rose

WHENEVER I CONTEMPLATE THIS ERA OF "only her hairdresser knows for sure," I regret the fact that my grandmother was born a generation too soon. Everyone knew for sure that Gram had never had red hair, but she had an obsessive fear of old age and took the only stand she knew against it. She was puritanical to the gray roots of that red hair, and nobody had greater disdain for "scarlet women" with their overpainted faces and their scandalous cigarettes. Yet she quietly squandered a small fortune on rouge and powder and perfume and, Heaven forgive me, color rinses for her hair.

That she should have feared growing old is paradoxical, for in her sixties she was as ingenuous as she had been at sixteen when she married and set about rearing a family in a log cabin. She was always pretty and had an educated pout that provided her with a fair share of her own way. She worked hard and did the things that were expected of a farm woman in her day—kept a great garden and a flock of chickens, worked quilts and wove rugs, and cooked mountains of food on her wood-burning stove. Nonetheless, she remained a wonderfully warm and uncomplicated romantic.

The daytime radio serials were her passion, and almost nothing was permitted to come between Grandma and her "stories." Her idea of a really good time was a lively exchange with someone as engrossed in "Our Gal Sunday" as she was. If she had foreseen a day when she could have sat breathless before a TV set and watched her "people" struggle through their episodes of misery, perhaps growing old might have held some small recompense.

The mark of youth that endeared Gram to us more than any other was her sweet tooth. A pinch of sugar went into everything she cooked, and we never had to twist her arm to persuade her to buy bags of orange slices and peppermint patties or a couple of pounds of chocolate marshmallow cookies. The Saturday trip to town to take the eggs was climaxed with sundaes or sodas at Newman's Drug Store, and nobody anticipated the treat more than Gram.

Another of Grandma's passions was the host of "true" magazines of love, romance, and scandal. As a young woman with broader vistas, my mother tried to "bring her up" a little, but finally despaired. When my sister and I were of impressionable age and Grandma was living with us, Mom quietly sneaked the magazines away and burned them (not always before we'd done a little sneaking of our own!). Gram scarcely realized how offensive they were to our wholesome household.

Purple was an old woman's color and she avoided it as she avoided sitting down thirteenth at the table. She surrounded herself instead with pink, which was youth. At a time when walnut stain was the going thing for kitchens, Grandma's woodwork was shocking pink. The fellowship of a Golden Age group would have been a perpetual quilting bee for Gram, but she never would have joined one. The Golden Age for Gram was the purple age, certainly nothing to be embraced. Twenty-five years after her death we still scrape layers of pink off drop-leaf tables, and memories of Gram float back on a pink cloud.

Reminiscence of Summer Long Ago

REMEMBER . . . SWAYING IN THE TOP OF A clump of willow along a pasture creek?
The banging of a screen door?
Bringing ice home on the bumper of a Model A?
Changing the pan under the icebox (or forgetting)?
A rooster crowing in the morning?
The "buck, buck, buck b'gawk" of a hen in a dusty hen house?
The keen pleasure of finding an egg in the nest?
Waiting for your brother to go away so you could ride his bicycle?
Smoking corn silk rolled in newspaper?
Learning the incredible facts of life in the neighbor's barn?
Mud between the toes and not minding?
The neighbors who complained to your mother that you picked their flowers?
Knocking at the neighbor's door and being admitted, and sitting silently watching their canary?
Reading Big-Little books on a porch swing?
The thrill of being allowed to ride in the rumble seat of your sister's boyfriend's car?
When a new oilcloth made the whole family happy?
Penny candy in a small bag, and not being able to decide?
When the cow got in the wild onions and ruined the milk?
When your cow went dry and you went to a neighbor's to buy milk, ladled into your jug from a big stone tub in the milkhouse?
Sitting through long funerals and watching fat people fan themselves, and not knowing exactly who died?
Waiting in a hot car for folks and honking the horn?
That the principal preoccupation of childhood was waiting, and that summers were endless.

Market Day

The market days of my childhood were always hatched from the cocoon of night several hours before the summer dawn. While mother went with a flashlight to feed the chickens and make last-minute additions to the load, we children dressed and ate our oatmeal. Then taking the change box and a crock of cottage cheese or butter from the kitchen, we made our way through the darkness to the truck and the long, sleepy trip to Cleveland.

Ours was a retail market assembled on an empty lot adjacent to a suburban shopping area; the hope was that before the sun and the customers appeared, we could create on our allotted space a commercial extension of our country garden.

The market load represented the total hectic effort of the previous day when the fruits and vegetables were picked or dug or pulled (whatever verb suited the object); sorted and packed and washed. Fruit was our "specialty," but like most truck gardeners of the era, we dabbled in everything from eggs to asters.

All this we handed down from the truck and spread about in traditional arrangement—potatoes on the left, corn at the side, flowers out front, culls to the rear, and so on. Alongside the change box on the market stand up front went quart measures of everything at higher prices. Down through the middle went half-bushels of apples and peaches in such quantities it seemed we never would sell them all. Then like a triumphal banner above the whole we raised the striped market umbrella.

When the setup was complete, we children were free to roam about the market until trade grew heavy and we were needed to fetch and carry and replenish the small measures. We were in fact encouraged to "run up the line" and take a quick price check on the competition.

Mama was a shrewd "market lady" with an air of confidence about her, and she usually commanded a price five or ten cents higher than anyone else. In my young mind that seemed justifiable; I always felt she was superior to the lot fate had dealt her of being a widowed farm wife with seven children. I very much hoped that her real quality showed through the babushka and butcher's apron she wore. (I needn't have worried!)

Market day was a wonderful mélange of smells: the odor of ripe melon as brother Henry (always our best salesman) cut a slice and proffered it to a reluctant customer—"There, just take a taste of that!"; the mingling fragrances of peach and plum; the tangy smell of peppers; the pantry smell of homemade butter. But most prominent of all, the heavenly scent of dill. I don't even remember that we grew it, but so pervasive was it through the marketplace that even today dill is "market" for me, as cotton candy is county fair.

There was a code of ethics about our market, understood if not written, that no one dealt in produce not his own. Middleman's profit was a thing of scorn, and a market "jockey" who bought wholesale and resold was the lowest form of merchant. There were some of whom it was said that their produce was "junk" bought at the downtown food terminal. (The irony is that for all of our farming life the Leimbachs have marketed their produce at that downtown food terminal and sold it to those market "jockeys.")

The market day wore on, the crowds grew thick, and the confusion was wonderful. Nothing thrilled us so much as being so busy we couldn't think, and emerging from the rush with a helter-skelter pocketful of dollar bills and change.

During those busy times we grew proficient at snapping the wormy tip off an ear of corn and slipping it in the sack when the housewife turned her back. (People who remember corn, infested as it once was with earworms, don't raise much fuss about insecticides.)

When the rush slowed in late morning there was time to exchange news with our market neighbors. We knew them well in that temporary environment, but we had never seen their farms. In our limited rural world, unless you knew the real substance of people you were inclined to be suspicious. Thus the market people had for us a strange unreality.

The climax of market day was having lunch in a restaurant and paying the check from that great wad of dollar bills. Then mother made the rounds of the candy store, the bookstore, and the department store, where she indulged us with little treasures we'd picked out by ourselves earlier in the day.

Taking down the umbrella finally, folding the stand, and bundling the empty baskets was welcome, but somehow anticlimactic. We were usually the last of the growers to leave. And as we pulled away from that empty lot in the heat of midafternoon, it seemed that the bustling market "village" of early morning had been some sort of dream.

Tea and Thee

OURS IS NO LONGER A "TEA" SORT OF WORLD, and I find that regrettable. It is, rather, a let's-be-up-and-doing coffee world with scant time or patience for tea. A poem on my kitchen wall that extols the virtues of tea summarizes, "tea is philosophy!" And this I suppose is the crux of the problem; it's a world that long ago tired of philosophy and hungers after fact.

Ours was a tea-drinking household principally because my father's family was English. The only time I remember my father really at ease was when he sat after dinner with his tea poured from the huge blue china pot.

When we drank our milk down a way in the glass we were permitted to have it laced with tea, and we naturally grew up in the tea ritual. Coffee was considered too stimulating a drink for growing children, so none of us learned to drink it at home. I remember how wicked I felt indulging in my first coffee in my second year of college. Thereafter I tried unceasingly to develop a coffee habit, but it has never really taken hold.

The tea memory I cherish most is of the late night pot shared around our battered kitchen table. Gathering from our separate pursuits, we found cups and saucers and pulled up chairs. Mellowed by the tea and the hour, we shared the opinions, the confidences, the doubts, the dreams that bound us, and still bind us. It was a communion you shared of your own volition, unlike supper or lunch when you were expected to be there. If you had gripes or hostilities you carried them off to a lonely bed. But if there was something of yourself that you would share, you tarried here over tea and toast. They were insignificant hours, yet perhaps in their way they were the most significant we knew, for out of them grew a security that allowed for confidence, where any fault was (and is) tolerated—over a soothing cup of tea.

Precious Quarry

THE PAIN OF PROGRESS IS MUCH DISCUSSED, but it is when a freeway cuts through the center of your own heart that you really feel the ache. For me it has been long in coming.

The densely wooded area that grew up around the East Quarries not far from my home was the paradise of my youth. In this huge wooded mountain of stone were at least half a dozen abandoned stone quarries that had filled with water and stood useless through the years. We called it simply "the quarry," but for the kids in our neighborhood it was Treasure Island, Sherwood Forest, and Never Never Land. There were caves and trails and shanties and a thousand secret places sought out and charted by a crowd of hardy little roughnecks.

That I was a female and not usually of the select who played there only compounded the enchantment of the quarry for me. It altered not a whit the claim of my heart on the place. I featured myself at least a Maid Marian to the fearless band. When my brothers sat at supper and discussed with bravado the battle campaigns of the day, or whispered at night the forbidden excesses of the place, I was enraptured. I became as much a part of their revelry as imagination could render.

On rare and splendid days I was permitted to tag along, but I was never allowed to forget that it was a boys' realm and I was an underprivileged visitor. I was skinny and small, but most of all I was foreign to the terrain. Stumbling along the rocky paths, up and down cliffs, in and out of underbrush I was always left behind. But never would I have complained, had the night come and found me bleeding to death. I never tired of being the "prisoner." That was, after all, my lot as the alien sex. My reward lay in the knowledge that during the supper hour I would play a real part in the exciting recapitulation.

One very glorious occasion stands out in my mind. On that day I was permitted to take my brother's twenty-two in hand and shoot at a soup can set on a ledge of the Little Quarry. I wouldn't swear now that I hit it, but in the 100 times I relived the moment, there was a hole clean through that can. And it seemed that there was new respect in my brothers' attitude. I remember that I was wearing my first blue jeans, which added greatly to my self-image.

The chief reason for the all-male quarry society was that bare-naked swimming was the rule of the place. My brothers enjoyed special status at the Big Quarry (the swimming quarry), being the most fearless of the high divers. It wasn't until I was fourteen or fifteen and had distinguished myself in their eyes as a pretty decent swimmer that they would throw their weight around in my behalf.

"OK, you guys, get your suits on. My sister's coming up,"

one would holler while I sat at a safe distance in a grove of sumach and waited for the "all clear."

In our secret souls, those of us who played there envisioned this vast hill of sandstone as an impregnable bastion from which on some illusive morrow we would stave off the world. We practiced its defense and yearned after the time when the mythical assailant would storm the walls and we would triumph valiantly.

Alas! The enemy attacked—twenty-five years later. A herd of bulldozers and earthmovers working day and night for two months plowed and dynamited a broad track through the East Quarries to lay down a bed for a superhighway. Not a shot was fired; no arrows or cannonballs flew from the turrets; no hot tar poured over the walls. The enemy entered and the kingdom fell. The Little Quarry is no more. The ledge from which I shot the can still stands, but no cavity is left for water; the trails and the caves and the secret places are gone. The Kingdom was Youth. Its defenders have fled, and yet . . . watch the cliff closely any day of the week and you'll see a motley band in patched overalls with faces reminiscent of the former troops, standing high watching the progress of the highway. These are our children who have inherited the Kingdom; they find it fascinating because progress belongs to it.

5 MYSELF

No book can teach us Self. It is a hidden language only Heart can read.
Joan Walsh Anglund *

The pursuit of wholeness is a constant journey into relationships. Thus it is that one is inevitably stumbling over his past in coming to terms with his present. The happy byproduct of reliving the keenly personal episodes and incidents of one's life is an acceptance, finally, of what one is.

An Episode of Roses

ORRIN WANTS TO KNOW "EVERYTHING ABOUT when I was a little girl," and I have tried to reconstruct it in episodes to please him, all but one shining day I always hoarded for myself, never knowing quite why, even though I remembered it so clearly. Finally one evening I took a chance on telling him. . . .

Mrs. Yaeger's rose garden was at the back of the park cemetery next to our farm. It was a sort of buffer zone between the burial grounds and the undeveloped wasteland beyond. It grows more grand in my memory than I'm sure it ever was in fact; where twenty-five or thirty rose bushes grew, hundreds seem now to bloom. It was hidden from public view by a high evergreen hedge, but was nevertheless carefully tended for the proprietor's wife by the cemetery gardeners. Knowing how painstakingly my mother worked over her flowers—zinnias, marigolds, and snapdragons grown among the vegetables—I had a tendency to resent the fact that Mrs. Yaeger only came to her garden for the harvest.

---

* From A *Cup of Sun* © 1967 by Joan Walsh Anglund. Reprinted by permission of Harcourt Brace Jovanovich, Inc., and Wm. Collins Sons & Co., Ltd., London.

If anyone fancied, however, that the beauty of this place was wasted on the sun and the sky, then they never noticed the timid little creature in bare feet and faded sundress who haunted the garden. For me that garden was more sacred than the burial areas. I spent hours walking the formal paths, imagining myself a grand lady who came in garden gloves to clip roses and lay them delicately in rounded wicker baskets. If anyone had asked me my favorite flower, I would have said an orange zinnia, but there was something about the delicacy of a rose that I yearned after.

Experience had taught me not to pick flowers in the cemetery and take them home in joyous handfuls to my mother. There had been plenty of deplorable incidents when Mr. Yaeger had driven into our yard in his big blue Buick and gently pleaded with my parents to "please do something about those children." I think the poor soul lived in dread of the day when one of our sacrilegious little faces would intrude itself among the floral pieces in the solemn midst of a funeral. No, we only took home the cut flowers in the wreaths three days removed from the graves, but that never stifled my urge to pick the live ones in the borders and the garden areas.

The day that gleams from the endless days of being a child is the day I picked all of Mrs. Yaeger's rosebuds. I had wandered over to the cemetery through the shaded glen where the creek bent to enter our valley. The water was running low over the stones, collecting in little pools, behind clusters of mossy rocks, trickling between them, bubbling over small rapids and into lower pools. No grown-up ever penetrated the overgrowth of willow that concealed this spot, so I went there often to play alone.

Up in the rose garden I was overwhelmed as always by the need to do something with all that beauty. The eye is an insufficient sensory organ for a child. He wants to come to grips with beauty—to eat it, clasp it to him, or give it in delight to someone. The vision came to me of flowers floating quietly down the creek through the glen, collecting in colorful pools at the lower end.

Like some mysterious wild creature, I gnawed off the buds shortly below the calyx, sparing none and collecting them in my skirt. I worked quickly and nervously like the thief I knew I was, and then fled to the glen on the safe side of the property line.

I carefully dammed the narrow channel where the creek emerged from the willows to flow through the pasture, and spent the afternoon floating rosebuds and creating water gardens. When the bell rang for supper, I left my roses corralled in several of the pools. In the morning when the buds were open it would be glorious.

But in the morning they were gone, every petal, bud, and blossom. They had flowed away when the water rose in the night.

"And then what?" said Orrin.

"Nothing—that's all," I said.

"What's so big about that?" he asked.

And I knew why I'd kept the story from him for so long. There was nothing "big" about it. The immediate effect of the afternoon, strangely, had been one of disappointment, so much so that I never forgot it. I had thought that having all those rosebuds would bring some sort of sensual satisfaction . . . but it didn't. And then I thought that perhaps when they all opened . . . but they never did.

Orrin would have found it a much better story if I had gotten a "good lickin'," if Mrs. Yaeger had acted on the anger she must have felt with this strange "fawn." I'm glad of course, that she didn't. Those stolen roses have bloomed for me all through the years. I learned much about beauty that day. But I can't explain it to Orrin . . . or to anyone.

Dance Perchance

It comes as a great surprise to me to learn that in some communities it is necessary to enroll a girl in the dance class of your choice years ahead to insure acceptance. The question in my mind is, "Why dancing classes at all?"

My dancing class memories explain my antipathy. Our class was held in the Odd Fellow's Hall up over one of the local taverns. The musty smell of the hall and of stale beer are my sharpest recollections of "dance class."

I remember the fear of mounting those stairs to confront this unknown terror, although it was all my own idea. Certainly nobody ever suggested to me that it would be nice if I considered "dance." I must have gone alone because my second strongest impression is that nobody ever took me seriously.

All along one wall were mothers who sat and smiled and knitted and chatted and acted pleased as punch with their little girls, but my mother was not among them. So, of course, I was jealous from the start.

I remember being told that dance shoes weren't really necessary, that you could just have toe and heel cleats put on your shoes, "But, of course, if you could afford shoes . . ." The cleats were much nicer anyway, because you could wear them around all the time if you only had one pair of shoes as I did and they made such nice clickety sounds in school and on the sidewalk.

The darling of the class (whose mother was always enthroned there with her knitting, her smile, and her younger daughters) had long blonde curly hair and looked far more like Shirley Temple than any of the rest of us. I really hated her and her mother and her knitting and her cute little sisters. It was an early manifestation of the perpetual fear that there isn't enough success to go around.

This little darling made remarkable progress. She had mastered the "shuffle, ball, change" maneuver in her lovely black slippers before I even got the cleats for my shoes. She went on to pirouettes and splits and who-knows-what-all. I certainly don't. I never got past that "shuffle, ball, and change."

No one at my house ever asked me to dance. They didn't

even acknowledge that I was taking lessons. It's possible that my mother was trying to keep the whole business a secret from my father. I doubt that he would have considered a quarter spent for dancing lessons as anything more than "damned nonsense!"

I don't recall any recital. If there was one, I was surely scratched as a poor contender. Maybe I didn't last long enough in the class. I don't remember quitting, but I suspect that the girl with the blonde curls came to be more than I could endure, and I quietly admitted to my mother that I didn't think I wanted to learn to tap dance after all. Anyway, the horror of climbing those stairs and "shuffle, ball, changing" before all those mothers went away along with the smell of the musty hall and the stale beer.

One of my nephews is starting dance lessons now and we talk about it. "I think it does something for them," his mother says. It probably does. Unfortunately, I've figured out what it did for me, and I wouldn't wish that on anybody!

Once on Easter . . .

It was a wool suit of blue and beige and white muted plaid, picked from the three or four available in Montgomery Ward's spring catalogue. It cost $12, which in the apple-orchard economy I understood very well for a girl of fourteen was twelve pecks of Jonathans or thirty gallons of cider. It was a considerable sum no matter how you figured it, and I was not ungrateful. My mother had bought herself a "good" dress for my father's funeral four years earlier, and I couldn't think of any since.

It seemed like some sort of miracle to have a real suit for Easter, a new suit all my own. If one miracle could happen, why not two or three? Perhaps I could have a hat and gloves—a whole outfit! Perhaps someone would send a corsage. Perhaps we would go to church on Easter. And for a wonderful anticlimax, I would have something new, garnished with a wilted corsage to wear to school on Easter Monday!

During all of the week before Easter I watched for an opportunity to bring reality to these dreams. A shopping trip for no other purpose than buying an unnecessary hat was not even considered; but I was working on a research paper that called for a couple of days of library work in town. One of my brothers would take me to Lorain by motorcycle in the morning and I would come home by bus in the afternoon.

By skimping a little on the research I got away early on the second afternoon and walked the long distance to the downtown shopping area. I remember the thrill of being alone for the first time on a feminine errand. At the same time I was haunted by a wicked sense of self-indulgence.

Where wars between ego and conscience were concerned, my ego didn't stand a chance; the visit to the little millinery shop was miserable. A beige hat was the "ultra" choice for that suit, so carefully checking the prices of two or three I chose a broad-brimmed felt with a grosgrain band at $2. It wasn't the sort of purchase that warranted a hatbox, but I clutched the bag they put it in and made my way quickly to the bus stop. The further indulgence of a pair of gloves was out of the question.

At home I furtively tried the hat with the suit for total effect, then hid the bag away in a closet where no one would ask about it.

On Saturday I went to a local greenhouse to help out as I was accustomed to doing Easter week. When I had finished the housework that was my detail there, I would sit in the workroom and watch the corsages being made. The names on the cellophane bags were often familiar ones; they were ordered by fathers for mothers and daughters, by boys for girl friends. If the family who ran the greenhouse had known my longing they would have sent one of the enchanting cellophane packages home for me. But they knew my family wasn't a church family, and they couldn't have suspected that I had a new suit and a secret hat. I'm sure they sensed that a corsage to wear nowhere would be unbearable.

It was late when I came home from the greenhouse, and I slept in on Sunday. If I had asked, someone would have taken me to church, but going alone wouldn't have fulfilled the dream.

I wore my suit to school on Monday, and when the subject of "outfits" arose I let it be known that I'd had a new hat too. No one needed to know that I hadn't worn it to church.

I never did wear that hat; I never believed in miracles again either.

Henrietta

IF I HAD DESCRIBED TO MY COLLEGE CONTEMPORARIES the little high school where I first taught (which I avoided doing), I would have called it a corny sort of place. As nearly as I could discern, the high points in the Henrietta year were the county basketball tournament in winter and the ice-cream social at the Methodist Church in summer. It was my impression that most of the young people hadn't been farther from home than Cleveland, forty miles east.

I had grown up on a farm but in a district that was much more suburban, and life deep in the country was something quite foreign.

If there was anything I was determined about that day as I stood trembling before the sophomore class in their blue jeans and 4-H cottons, it was that they wouldn't make a hick out of me! When I had gained a little experience I was going overseas to teach for the State Department, and goodbye to all this.

I was about one quarter of the high-school faculty that fall—the English, speech, modern language, and girls' phys.

ed. departments. I coached the girls' baseball team and the cheerleading squad, directed the plays, supervised the paper and the yearbook, advised the G.A.A. and the sophomore class—all for $1900.

Throughout the fall there was a dance or a party to chaperone nearly every weekend, and after Thanksgiving the social season was all basketball. By midwinter I was anticipating the county tournament just like the natives. I will never forget the delight of showing up for the ice-cream social that following August and being welcomed with hugs and squeals.

These country people wrapped their hearts around me and claimed me in spite of myself. They introduced me to hayrides and square dancing and all the unmarried farmers. When I married Paul their claim on me was complete; I became a kissing cousin to half of them. The truth was, of course, that I was a born "hick"; I just hadn't found myself until I came to teach in that country high school.

When my sophomores were seniors they asked Paul and me to chaperone their senior trip to Washington and New York. I remember well the evening we sat in the Latin Quarter in New York City watching a very nude stage show (part of the package tour); I looked at all those ingenuous faces and asked myself what in Sam Hill we were doing in that place with those kids from the Methodist Sunday School. All in all they were unimpressed with the big cities—"nice places to visit" and all that. They missed their square dances and the local picture shows.

Some of my students from '49 are scattered over the country and the world now, but a fair share of them settled here in this rural area. I recognize the names and the faces of their children when I go to substitute teach in the consolidated system that took in Henrietta. When we "hicks" get together at reunions and parties and PTA carnivals, we all agree that this is a great place to live . . . but I don't suppose that anyone would want to visit here.

One Woman's Walden

One of the fringe benefits of my life is that I get to travel a lot—driving a truck. On an errand to the basket factory one summer day, I sat on a crate and worked at my writing while the men loaded the truck.

"Whatcha' doin' there?" asked one of them.

"Oh, I'm working on an article," I said.

He thought about that, then replied, "It don't make sense, a truck driver writin' an article." He was implying, I suppose, that truck drivers have less acumen than people otherwise employed, or perhaps that driving a truck is something less than stimulating. But I am quite beyond insult and certainly not in agreement on the point of stimulus.

A farmer's wife who drives a truck travels "incognito," and this is novel. She explores different worlds—the back rooms of grocery stores, the kitchens of restaurants and schools, the grain mills, the fertilizer plant, the bag factory, commission houses, roadside markets, truck stops, and many more. In addition to being beneficial body-building work, truck driving is pleasant and frequently amusing.

Turnpike ticket takers tell you that truck drivers are getting better looking every day. Other truck drivers do a double take when you pull up at a stoplight. Pedestrians along the street drop their jaws. Little kids sitting on fences are delighted if you blow your horn for them. Merchants view you with alarm when you pull into their crowded alleys, and in the end admire you for your maneuvering skill. The few women who don't consider you insane will praise you for your courage. It is very easy, in fact, to get an inflated opinion of yourself, unless you make the mistake of taking along a teenage son.

He will proceed to rip your driving to shreds and otherwise convince you that as a truck driver you're pretty much of a "retard." You don't go fast enough, you're in the wrong lane, you're giving the other truck drivers the wrong signals, you're taking the wrong routes, your shifting leaves something to be desired. And worst of all, you're all wrong in the operation of a two-speed rear end.

A farmer with the temerity to send his wife on the highway for such a pursuit must, of course, recognize the liability involved. It will cost him a bit more in gasoline, for she's bound to be lost at least part of the time. He throws her on the mercy of other men for things like latching the back door and putting up the tailgate. In a flat-tire crisis she's no good at all. In the first place she can't lift the jack, and if she could manage that she couldn't lift the tire.

She copes quite well with running out of gas, having had more experience at that than he'll ever have. There's a helplessness about a woman standing with a big truck along the highway that is irresistible to the knights of the road. Men are so quick to be helpful that if a woman stops intuitively to check for a flat or chase down a "something burning" smell, she does so in a manner indicating that she's inspecting someone's abandoned property.

If she takes her small children with her, they enjoy hours of uninterrupted conversation—time for folklore and language practice, for memorizing songs and rhymes, for joking and observing, exploring and learning.

If she is alone she has long stretches for contemplation and searching the mind. The lonely night hours have a wonderful mystic quality on the highway when reminiscence is keen and introspection a revelation. In a sleepy mood she turns into a met soprano. I sing beautifully to myself at night in French and German; patriotic songs and lusty alma maters—music that only I want to hear, but I love it.

It's probably true, what the man at the basket factory said: "It don't make sense, a truck driver writing an article." But behind the wheel I am Thoreau! . . . and I have friends who say my driving looks like it!

Moon, June, Loon

SUMMER MOONLIGHT IS A WELL-KEPT SECRET; the world at large supposes that it belongs exclusively to the young in love. If you have ever been young and in love and have now reached an age to think rationally about it, you should realize that lovers have small concern for the moonlight. It may be some sort of catalyst, but enjoyed for its own quiet delight? Huh-uh. The young in love enjoy only each other, moonlight or no moonlight.

To fully appreciate the phenomenon of moonlight you must venture into it alone, or at least at a stage in your relationship when you can be "alone together." When I head out the back door intent on capturing a moonlight mood and observe that the doorknob is sticky, "moon-June" may still rhyme with "spoon," but the spoon has been dabbling in syrupy Kool-Aid. I turn back to the sink for a dishcloth and there's no question about it, the honeymoon's over!

Five minutes on the porch moon gazing and you know why moonlight goes basically unexplored. There is a self-conscious feeling that unseen eyes are watching and judging you for some kind of a nut. But reasoning that my kin are sound asleep and my neighbors out of range, I persist. It begins to reach me, the concealing, revealing magic of the light of the moon.

The barns, strangely, are the loveliest sight; strangely, because by day they are an eyesore in their varying states of painting and disrepair. But a profile of barns by moonlight, all angles and planes in tones of charcoal, is beautiful.

The flower garden too is a different sight by moonlight. There should be roses, I suppose—the refrain of "Moonlight and Roses" drifts through my head. But having acknowledged that my motive is moonlight and not love, who needs them? I suspect that the people who have success with roses go to bed at a decent hour so they can be up early dusting. I doubt that they trifle with moonlight. There is no need to observe at a nocturnal hour that the orchard grass is growing thick in the iris, or that the bergamot is crowding out the primrose. Only the long spikes of delphinium and the red-hot pokers tower over the mass, and they seem to set it all in order.

Turning my back on the garden and the barns, I look at the house as the moon sees it. Even the poor dying apple tree that shades my kitchen looks lovely by moonlight.

I think of all the moonlight clichés and recognize that poets as well as lovers have a great claim on moonlight. They have almost worn out the words. Describing it even for myself is futile. The scent of honeysuckle is strong in the air as I go reluctantly back to the house. The bees and the butterflies and the hummingbirds have left it to me for the night, so in the half light I pick a handful, trying to choose the still-budding stalks, that they may give me a moonlight memory a little longer.

After all, a vase of honeysuckle makes a very nice poem. I put it on the bed table where Paul will wake to find it in the early morning. Perhaps then he will forgive himself for marrying a dreamer who wanders about in the moonlight instead of getting up to dust roses at dawn.

Summer Dawn

ALTHOUGH I MUST FREQUENTLY BE UP AND about with the dawn, I am not by nature a "morning type." (I say it's low blood pressure; Paul says it's bad habit.) Ask anyone who normally gets up before the rooster crows and he will tell you without reservation that dawn is the greatest time of day.

The dawn is an evolution of sensory perceptions—the

gathering light, the shapes emerging, the cessation of the night sounds, the sudden wakening twitter of birds, the freshening of the breeze, mist rising from the valleys, the mingling of odors made heavy by the moisture—emerging awarenesses becoming tangible day; the infinite becoming once more finite. Dawn is the surrender of the privacy of night to the community of day. Dawn is for everyone who experiences it a very personal emotion.

Summer dawn for me never loses the original magic of my discovery of it. Dawn is my brother John whispering in the dark, "Wake up! It's almost daylight. . . ." And we were instantly awake, brother Bill and I, climbing into our waiting clothes, stumbling down the stairway in the darkness to the kitchen where Johnny was already toasting bread in the oven. We would pull on our shoes, butter our toast, and holding it in our teeth, put on our sweaters and file out the back door.

Money, of course, was the motivation. Pocket money was difficult to come by, and we seized upon every scheme we thought could produce a little of it. City women would pay as much as twenty-five cents a quart for blackberries, and that was *big* money when I was nine or ten. But we had to roam far afield to find enough berries to fill the demand. Hence the sorties before dawn.

We sometimes walked two or three miles to reach berry bushes we had scouted on previous days. Strangers often took a dim view of kids picking their berries, so we aimed to finish and be gone before they were stirring.

We talked very little—silence was part of the required stoicism—but there was great unity in our silence, and it is probably that silent camaraderie that reaches me down through the years in the privacy of dawn.

The picking was also done quickly and silently, but the return trip was chatty and fun. There was a swagger in our manner that clearly indicated we'd been into mischief. When we hit the house where the rest of the family was only beginning the day, we were boisterous and loud, and to everyone else's annoyance, full of our adventure. All through the day we invented awkward little ways of telling how early we'd been up and how much we'd accomplished.

There were other junkets at dawn—hunting bittersweet in October, checking trap lines in winter. Always the ritual was the same, John in the lead, Bill following, and Patsy, of course, trailing.

Dawn has never lost for me that sense of urgency and excitement born in those childhood experiences. To know and savor a day to the fullest is to know it at dawn when it is only form without detail and color—those of us who customarily arise after the day has bloomed have surely missed the perfect promise of the "bud."

Literary Landscapes

THERE IS A VERY SPECIAL SORT OF YOUNG GIRL who will pass the summer oblivious to heat and household routine, picnics and pool parties, vacation and vexation. She is the asocial creature suspended in the stage between baseball and boys, whose all-consuming passion is books.

She eats her cornflakes with a book, washes dishes in the shadow of a book, and spends the balance of the day sprawled sideways in an armchair with a book. At night she kicks her jeans across the base of her bedroom door to conceal the fact that she's reading in bed. Hauled off on vacation, she will look up from her book long enough to remark, "Oh, are those the Grand Tetons?" It's an insufferable stage; I'd love to live it over again.

I have always promoted literature as a vehicle that transports the individual into all the world. In reality, nothing I ever read took me more than a few miles from home. Literature only succeeded in bringing the whole world to me. On the limited landscape of my neighborhood, my teenage mind created a permanent stage set for the characters of all the world's books.

Bluebeard's wives watched fearfully from the upstairs windows of the barn. The Little Shepherd of Kingdom Come roamed through our peach orchard. The Civil War as I reconstructed it from many novels was always fought in the north end of the old apple orchard. Christian's progress to the Celestial City was through the plum orchard, along the rocky bend of the creek, and up the hill beyond.

Lorna Doone's Scottish hills were above the creek valley just behind the machine shop, and just beyond the creek rose Tom Sawyer's island. By some strange alchemy the old corncrib at Beetlers became the ivy-covered walls of Oxford, and just outside its door at other times played Penrod and Sam.

Whether the book was *Vanity Fair* or *Les Misérables*, the Battle of Waterloo took place in the side yard of the Croyle home next to Beetlers. On the slope behind that house stood the slave cabin of Uncle Tom within a stone's throw of the fateful field where Bonnie Blue was thrown from her pony and died.

Most of Poe's chilling stories took place in the Croyle house. "The Telltale Heart" beat from beneath the floorboards of the northeast bedroom, in the very room where Raskolnikov much later killed the old lady in *Crime and Punishment*.

The Bliss house in Avon had an open staircase, a parlor with fireplace, a pantry, and a rear stairs. For me it was Tara, Manderley, Wuthering Heights, and an endless succession of lavish dwellings. It was surrounded by the English moors where Heathcliff roamed, where highwaymen ambushed stagecoaches and Sherlock Holmes hunted the Hound of the Baskervilles.

David Copperfield was born in my grandmother's kitchen, Cyrano de Bergerac played his great bakery scene in Thomas's bakery in Amherst—Greece, Rome, the Orient, and ancient Egypt all had specific locations in that crowded territory. No matter how my literary horizons expand, they are confined to the physical setting of my childhood between ten and sixteen. I have only to close my eyes and ponder, and I can direct you to any one of a thousand fictional spots.

So don't intrude where little girls sit with their books and dream. They are busy creating landscapes, shaping settings to which they can retreat at will forever afterward. And though they may seem to be off on another planet, chances are good that they are only orbiting the backyard next door.

Bargain Offer

I THOUGHT SHE WAS A POTATO CUSTOMER WHEN I let her in, but it turned out that I was the customer, she the merchant.

What she was selling was "Eternal Life" in a blue leatherette cover for $1.50, but I didn't buy any.

"How would you like to know that you would never die?" she asked.

"Well, I don't think I shall," I said.

"You mean you don't think this body will die?" she said.

"I would hate to think I were confined to this body," I said, glancing in the mirror thinking what a sad sack of bones it would be in another forty-five years. Thinking also how poor I should be if my spirit couldn't wander back to places I'd been—to the castle window where I sat looking up and down the Danube Valley, to the meadows of Czechoslovakia; or to places where I have never been, save in spirit—to the high, cruel coastline of Big Sur, to little squatter shacks in the Mississippi Delta, to Arab tents in Yemen, to bleak, isolated wood homes in the Tundra—all places I have rejoiced, suffered, or relaxed placidly with other human spirits.

"I don't think the important part of me will die," I said again.

But that didn't satisfy her. She was talking about the physical body living forever. She brought up Methuselah, and I thought of all those stories of people misplaced in time, how disoriented they were—Rip Van Winkle, Buck Rogers, Billy Pilgrim from *Slaughterhouse Five*.

No, I don't want any, I insisted.

The sales approach says you must create a desire for the product, so she tried a new pitch.

"Don't you believe the Bible promises us that we shall live forever?" She drew out a small Bible that relaxed comfortably in her hand. She thumbed with great familiarity through the much-underlined pages.

"But you see, I don't believe everything it says in that Bible just the way you believe it," I said. I didn't really want to shock her, but in any debate I knew that a few things must be understood. In this case, our differing interpretation of the Bible was one of them.

She read me a few passages anyway, concerning suffering and death. "Don't you believe we should try to eliminate suffering?" she asked dubiously, as though I were some sort of lunatic she hadn't encountered before.

"Well, yes, of course," I said, qualifying my remark. "But, on the other hand, if there were no suffering there would be no deliverance. If there were no sorrow, would there be any rejoicing? I'm not sure. We understand the one in terms of the other. And by the same token, don't we define life in terms of death?"

How could I explain to her what I felt about balance in life, the need for everything to fit together for the whole of life; black, white; day, night; male, female; good, evil —the Yin and the Yang.

I thought specifically about death. I thought of the perilous moments in the previous week when I stood over the unconscious body of my wounded son and asked myself, "What if he doesn't live?" And I told myself over and over, "He is a precious, precious part of your life, but he is not all of it. There is much, much more."

I look at him now, young, and in a mother's eyes, beautiful, and even more precious than before.

"What I really believe," I said to the woman, "is that life is so rich, it's presumptuous of man to ask for more than it gives."

Eternal life? Not for this body. It's a good enough body, but much too confining. I want to float with the birds and blow with the winds, open with the leaves in spring, rustle through the summer, drop in the autumn. I want to fall with the snowflakes and melt on the mountain tops and flow with the rivers to the sea. If I may. But if that is not part of the whole, then I shall have been content with what is and was.

In the end she wanted to give me "eternal life."

"No thanks," I said. "Save it for someone who needs it."

Born Loser

FEW OF THE TRAUMATIC MOMENTS IN MY LIFE have been so horrifying in themselves as in the obligation they forced upon me of confessing to someone that I had done something stupid. There was, for example, the time back in tire-rationing days when I drove two miles on a flat tire (it was harder than a son-of-a-gun to keep that truck on the road!) and then had to go home and face the fire flaring from my brother's nostrils.

And there were all those frightening conversations that began, "Honey, I've had a little accident . . ."

So on the day I lost an $894 check endorsed for deposit, I broke into a cold sweat at the prospect of calling home from the shopping center to say, "Honey, you know that check . . ."

I had started my errands, I knew, at the lost-and-found office of one department store, where I inquired to no avail for a pair of gloves lost on a previous excursion (third pair that year). From there I went to the dime store to make sundry purchases, then proceeded to a second department store where I bought a garden spade.

It was at the peanut counter while grappling with the spade in my search for loose change that I noticed I'd lost my small packages. I rushed back to spades, where three idle salesmen happily reunited me with my parcels. It was then I discovered that my bank deposit envelope was gone.

Offering hysterical little entreaties to the deity, I made frantic inquiry there, then back to the peanut counter. "Have you seen . . . ? No."

Back to the dime store, eyes searching the route, mind picturing a sinister character with shifty eyes and turned-up collar cashing my check somewhere that very moment.

The checkout clerk at Woolworths was a grandmotherly woman of the sort who might once have gone door-to-door measuring ladies for corsets, achieving a sort of intimacy to which she was not entitled.

Panting from haste and fright, I asked, "Have you seen a white envelope . . . ?"

She fixed me with a "you naughty girl" look and said, "What was in it?"

This was precisely the sort of judgment I dreaded.

She doesn't have it, I thought in panic, or she wouldn't torment me like this. I hesitated.

"Uh-huh, a check," she said. "A big check—for eight hundred and some dollars—right? Endorsed."

"Right," I said weakly.

I was eight years old: I'd just lost my milk money, and I was about to be treated to a resounding lecture on carelessness. But she didn't deliver it. She just held me with that reprimanding stare while I relived the accumulated scoldings of forty years.

"It's already in the bank," she said flatly, and with that she released and forgave me as too many people have in the past. "I gave it to the manager and he took it down to the walk-up window."

She didn't want my thanks, as grandmothers seldom do. It's their mission in life to look after careless little girls. And no one grows up so persistently confirmed in the charity of people as careless little girls.

Physical—or Was It Mental?

THERE MIGHT BE A CERTAIN MARTYRDOM IN having it said of me when I'm gone, "Poor girl! She neglected herself so," implying all-encompassing devotion to something higher than self. The posture, unfortunately, doesn't suit me; so biennially I make an inventory of all my ills and haul them to the clinic for a checkup.

Presenting my specimen bottle, discreetly concealed in a small sack (which looks like nothing so much as a small sack discreetly concealing a specimen bottle), I am admitted to the inner offices, then shunted off to a lab for blood tests. I sit toe to toe with a young woman who is nervously studying the bare fittings of the tiny waiting room, and when her eye catches mine I feign absorption in a large blue-white-black photograph of a glacier that commands one wall. I wonder idly why anything so stark was chosen for a place where warmth is so wanting.

When the bloodletting is over it's back to an examination room. I am instructed by the nurse to strip and put on "this gown," which resembles two large paper napkins crimped together with an aperture for head and neck and stretches to just above the navel. I swathe my remaining self in a huge paper sheet and stretch out on the examining table trying to pretend it's someone else lying naked under this great waste of paper toweling.

The doctor enters, relieving me from preoccupation with my cold feet. He queries me at length about my medical history. We talk about my relatives, some of whom were dead before I was born (and how should I know what they died of?). Medically speaking, my family sound like a very dull lot—no T.B., no cancer, no diabetes, no heart cases . . . one gets the impression that only murder or suicide would carry them off.

Then we get down to present problems, talking clinically and dispassionately about the body under the paper sheet. The examination begins—the thumping and the poking, the deep breathing and saying "ah," the whole self-conscious routine. Intermittently the salient facts about this "foreign" body are recorded on tape, and I have the feeling that eventually it may be programmed into a computer. Statistically I'm in great shape.

When it's over and he's gone, I shed the paper trappings, and the two members, head and body, are one again. I dress with new appreciation for the warmth of clothes.

"Now, tell me what you worry about," says the good doctor when I am seated in a leather armchair opposite him in his business office for the final civilized conversation, without the alien presence of that body under the sheet.

"Gee, I don't worry about anything," I stammer.

"Oh, but everybody worries about three things," he says.

"No," I said. "I don't. Things are fine with me."

"Do you worry about the school?" he asks.

"No, we have good schools."

"Do you worry about your husband?"

"My husband's fine. I don't worry about him."

"Your mother? Your children? Money? Your brothers and sisters?"

"Nothing." I begin to feel that I am flunking the oral exam. "Of course, I worry about the state of the world, but I don't think it helps to lose sleep over it."

"I guess we all worry about that," he says, dismissing this as a root cause of anything. "You know life is like a pendulum, swinging high for us, then low."

"I don't know where I went wrong. I only seem to have highs."

"I've got news for you," he says, and we part on that note.

I have the conviction that he was talking of somebody

else—maybe that body under the sheet. At least now I have some worries. What are those three things everyone's supposed to worry about? I never did ask.

I glance at the bill the receptionist has handed me; I have the feeling that the pendulum is on its way back. . . .

First Day

I'M PREPARED TO AFFIRM THAT A LITTLE knowledge is a dangerous thing, as from my kitchen I follow the conversation on the stoop outside. Orrin and Lynn are tormenting their younger buddy with the accumulated learning of a first morning at kindergarten:

"Dennis, you wouldn't know how to get in line."

"And you'd probably talk without raising your hand."

"And you don't know how tall the slide is. It's a lot taller than your slide, and about this wide, isn't it, Lynn?"

Poor Den. He was prepared to be chagrined when his constant companions went off to kindergarten, but the vastness of his ignorance was a threat he hadn't anticipated.

"And you don't know the rule about slides either."

"Or the swings rule!"

Having thus exhausted what they saw as one day's lessons, they go off seeking other ways to prolong their new sense of togetherness.

Watching them, I wonder just how much things have really changed. To that inane first-day question of, "Well, what did you learn at school?" my answers would have been similar to theirs: "My teacher is pretty," I would say. "She has red hair. I forget her name. Boys are supposed to take their hats off in school. We marched downstairs to go to the bathroom. We had two recesses and lunch hour. There's a thing called a fire escape by the window. . . ." But what I really learned hurt so deeply that I couldn't have spoken of it. Instead, I hid it away with all the secret aches that bear on what one ultimately becomes.

Because I had been preceded to my first-grade room by five brothers and sisters, it was not deemed necessary that Mother bring me on the first day. I was sent instead with my sister. I certainly wasn't the only one there without "Mama," but the presence of even one or two parents gave a timid child the feeling that he had been somehow neglected. The only familiar person in the room was a boy who lived across the cemetery from my home. I had seen him once or twice at a distance, but we had nothing to talk of together and could not have passed for friends.

And so I sat nervously picking at my nose, fidgeting with a strand of my very straight hair, and envying the other children, all of whom I fancied were friends and "belonged" here where I so obviously did not. So reluctant was I to meet anyone's gaze that I might not have noticed a boy was laughing at me had it not been for the admonishing remark of a little girl who, I jealously observed, was his close friend.

"Don't laugh at her. She doesn't know any better!"

If I still harbored hopes of the bright new world of school, they were shattered in this first rude lesson; I had learned much more than that it is gauche to pick your nose—I didn't *know* any better. Everyone else did. Clearly the world was unfriendly, much more unfriendly than I had imagined.

It was five-year-old logic, and valid as such. The two "friends" were with me through all my public-school years. I somehow outdistanced them as scholars (when you're number thirty-two you try harder!) but I don't think I forgave them until today, when I listened to these two little friends on my back steps. What they really taught me, beyond those early misconceptions, is that life is always easier to face when you are fortified by a friend.

Thanksgiving 1946

On my nephew's twentieth birthday I mailed him a note of congratulations and indulged a humorous memory of the week he was born. In the mid-1940s a bridegroom did not go to work at a salary exceeding his father's, and the practice of living with the folks was still a common thing. A young couple often lived in until they were "three" or "four," to the discomfort of everyone involved. Michael was my sister's second child, and he was born the week before Thanksgiving. Number one was a whimpering one-year-old caught in a cross fire between the toilet-training notions of her mother and her grandmother.

Aunt Pat was a college girl who had just performed the coup of the fall semester by landing a date for the Thanksgiving Day game. The lucky boy was pompous and egotistical, and the aura of puberty hung heavily upon him. He compensated for the latter characteristic by wearing a hat, which only succeeded in making him look totally inane. We had been thrown together at one of those exercises in prolonged embarrassment known as a dorm-frat exchange dinner, strategically timed to fall two evenings before Thanksgiving. Considerations of charm notwithstanding, there was a lot of squealing in our corridor when I announced that I had a date for the game. (I was no queen either!)

So Thanksgiving morning Prince Charmless and I went to the big game. The contest itself is as blurred in my mind now as it was then. I *do* remember that he didn't buy me a football mum, mumbling something about didn't I think they were a ridiculous extravagance. I wasn't about to tell him that I thought a chrysanthemum was the only really good reason for going to the Thanksgiving Day game.

We lost the game, it started to drizzle, and everything

palled. The drive home was interminable; conversation an agony. He was a mother's boy, I gathered, and wealthy (and no chrysanthemum!). "Mother" had taken him to Europe "before the war," when my mother, I reflected, was struggling to keep an anonymous villain from foreclosing the mortgage. I was immediately defensive and wracked my brain for some topic that wouldn't reveal the fact.

"What are your plans for Thanksgiving dinner?" I asked as casually as possible.

"Well, I thought I might be invited to dine on turkey in the country."

I nearly bit off the thumb I'd been chewing, and stammering through a profusion of "Gee's" and "Golly's" I rapidly explained that my sister had just come home from the hospital with a baby and, "Uh-uh, gee . . . well . . ." I just didn't know what was planned at my house. I had visions of his sitting there at our table in that dreadful hat, my brother-in-law and my brothers scrutinizing him and judging him not nearly good enough for Pat, and their not realizing that he'd been to Europe before the war and was far too good for me, and, oh, the whole thing was unthinkable.

A few more miles of static silence and we arrived at the farm. There is one thing of which a college student is certain: The house will be in perfect order when he comes home. He is "company" now, and homecoming is a special occasion.

Charmless walked me to the door, and stood there! Since I had no intention of kissing him goodbye, I did the only civil thing I could think of: I invited him in to meet my mother.

The place looked like something out of *Grapes of Wrath*. My sister lay dying (apparently) on the couch. My brother-in-law knelt helplessly by, holding her hand. Offspring Number one, training pants drooping, wailed forlornly; and a crimson Michael screamed from the bassinet. The dining room looked like a flea market. It was three miles from the front door to the ironing board in the kitchen where I found my mother. The introduction was as rapid as if it had been conducted on an escalator, and Charmless was gone!

Somewhere in a restaurant that evening he and his mother had their turkey, and as they pondered the season, the harvest, and rural America that made Thanksgiving the feast it is, I can imagine that he shook his head and said, "But you wouldn't believe, Mother, how those peasants live!"

Substitutes Are Safe— Not Always Sane

THE LOT OF THE SUBSTITUTE TEACHER HAS never been an easy one. Even when I was a student, there was a great whoop in the fifth-grade room when it became known that Miss Ehrman was ill and we could look forward to a day of free-for-all with a substitute. Few people survive long in the business, usually capitulating to a regular teaching assignment where the pay is more and the harassment less.

But I have made something of a career of substituting and learned many of the tricks of the trade. I know, for example, that it's best not to call on students from the seating chart. They delight in changing seats and answering to one another's names, usually good for causing a small riot. I have learned to look through the grade book and pick out the kid who'll give you the straight scoop on the assignment, and the wisdom of evicting a "catalyst" to settle the "brew." I have learned to "tread water" in almost everything from Julius Caesar to soccer, and how to size up the capability of the regular teacher, the administration, and the school board.

I have operated on the theory that if you go into a classroom, "hang loose," and assume that you'll have order, you'll probably get it. And through twenty years of the ordeal, I have managed to remain reasonably sane and relatively lovable.

But a solid week of monitoring study halls for six hours daily has left me with a bruised and battered psyche. I have had to remind myself every hour on the hour that my mother loves me, my children commend me to their friends, my brothers and sisters once sent me a card saying, "You're the greatest!"; that my husband looked around a number of years to find me, the dogs wag their tails when I come home, and I still have a few friends (who haven't any children in these study halls).

Survival technique is to get up in the morning, look in the mirror, and say three times with conviction, "You *are* a human being!" And then keep muttering it to yourself all day when the kid on the south side of the third table comes up for the eighteenth time to ask if he can "get a drink . . . go to the restroom . . . talk to Willie Winsome . . . go to the office to find a lost wallet . . . go make a phone call . . . go to the gym and jog . . . go out and take down the flag . . ." And then he climaxes the day by asking why you give him a detention for crumpling a wad of paper and aiming it at the basket. "Why are you picking on *me*?"

The usual teaching career must crest at about twenty years when a teacher approaches mastery of his subject, his caseload has been reduced somewhat, his salary increased, his seniority established, his authority unchallenged. The students he agonized over in his green years have achieved a measure of success and send him their children forearmed with respect.

But what of the career substitute who wanders in limbo through the years between a variety of schools and subjects? I don't know where or when her career crests; but on this January Friday as I speed homeward on the freeway in a compulsive release from the repression of thirty hours of study halls, I have the feeling that mine is cresting.

One of the best-selling books of recent years is a thoughtful analysis of achievement called *The Peter Principle*, Peter's principle being that every person succeeds finally in rising to the level of his greatest incompetence. It has taken me twenty years, but I think I have arrived.

Perpetual Valentine

I USED TO WONDER A LOT ABOUT THE *Farmers' Almanac* and its reasonably accurate predictions of weather months and years ahead. It has since become clear to me that by plotting on map and calendar the farmers' conventions in all the states, it's a minor matter to determine at least when and where the most intensive cold spells of the year will occur. Those are the periods when the farmer's wife finds herself in charge, and no matter how busy she's been all year telling people how she runs things, the truth emerges. . . .

"I don't know what you're worried about," says Paul as he puts on his topcoat. "The furnace is burning fine. Throw in a little coal at noon and chore time and shut 'er tight at night. The chores will take you ten minutes. Flip on the pump switch, throw out four bales of hay and three tubs of feed—simple as that. Any trouble, just call me." And he's off, as I stand there trembling in my flannel nightgown.

As the car disappears over the knoll, the snow begins. The afternoon is blizzard, and by suppertime the temperature has dropped twenty degrees. The well has gone dry following a day of washing, and water must be carried from the barn. "Ten minutes" of chores takes the boys and me forty-five with no special problems. Snow and wind continue, and the temperature falls to minus five degrees. By getting up in the night and stoking the furnace I manage to get the house temperature up to sixty—something about drafts and dampers I haven't learned yet.

Morning is something else. I get the boys off to school and go out to my chores. The pump is frozen as well as the hose, and I take it to the furnace room, get an electric heater for the well pump, and while it thaws I chase down a sledgehammer to break the ice in the water trough. Stopping by the potato basement, I find the temperature falling there, so I lug some kerosene and light a few stoves. All the while I'm plodding around the barnyard, well-dressed ladies are arriving at the house where I'm supposed to be hostess to a church meeting. The minister pulls in just as I'm carrying out my thawed hose.

"Anything I can do to help?" he asks.

Assuring him that seminary never prepared him to cope with a farm on a subzero morning, I send him in to coffee, where fifteen shivering women huddle for warmth and are ready with suggestions on how to fire a furnace. Because nothing will keep a northwest wind from whipping in around these old door frames and window sashes, we crowd into the dining room on the southeast and make light of it all.

As the last guest escapes, frozen stiff, a gust of wind rips back the storm door and it bangs against the house. Bundled once more in barnyard clothes, I struggle with the door, using a screwdriver unearthed from the Erector set. A phone call interrupts to tell me that somewhere there's a boxcar of fertilizer I should be unloading! Back at the door I reflect on my situation . . . "If you need me just call." Things could be worse, I think, remembering previous convention weeks: when the steers got out, and the year the baby arrived a month early.

During the afternoon the wind subsides but the cold persists. At least the house is warmer. Temperatures are still at dangerous levels in the potato basement and we're out of kerosene. When the boys come from school we get a tractor started and tow the truck which balks with the cold. After we've done our chores we go in search of fuel. The woman at the pump of the local gas station looks at me as though I'm a little demented. I don't envy her either, and we pass a few words about the folly of women who even dream of leaving their husbands.

Driving home I think of all the cold, bitter, frustrating experiences that have been this day, of how I'd like to vent my spleen on somebody. The phone is ringing as I come in the back door, and still bundled in barn clothes and boots I scurry to answer it.

"Honey?" says the voice, "How's it all going?"

The bitter thoughts have spent themselves under the cold moon, and there is only one truth. "Have I ever told you how very much I love you, and how I need you—desperately?"

Instant Psychedelic

THE DRUG SCENE AND ITS PASSION FOR TOTAL escape is a frightening thing for the parent generations. Perhaps there is a validity to the viewpoint that life cannot always be lived and endured on the level of reality. But my generation and the other "deprived" generations have somehow learned to survive on what might be called "short trips." Each of us has a personal catalogue of satisfactions that serve us.

I am transported by the mingling aromas of coffee, smoke, perfume, and good food cooking. I love the look of a three-year-old's perfect little body, the sound of a Mozart concerto, the feel of a flannel nightgown when I'm chilled and exhausted and flirting with illness.

I am delighted by the perfect nonsense of a Dr. Seuss —"Did I ever tell you that Mrs. McCave had twenty-three sons and she named them all Dave?" * And I am elated with a thought so beautiful it must be felt to be understood—"Keep a green bough in your heart, and God will send a singing bird."

There's a tantalizing hint of sun and shadow that strikes my sink between ten and twelve o'clock in the morning

---

* Dr. Seuss, "Too Many Daves," in *The Sneetches and Other Stories* (New York: Random House, 1961).

and makes me sad for people huddling in darkened rooms getting stoned. Then there's always flower power—I can stare for hours at a bouquet of poppies; and bluets blooming on a barren hillside are unspeakable joy. I have sat up late feasting my eyes on a rose whose beauty I knew would not survive till morning.

Going into a quiet room as the benediction of a day and pulling the blankets around your sleeping children is a mute celebration unequaled in human experience.

These are a few of the "short trips" on which I have ultimately found myself—or "lost" myself, I've never been sure which. Perhaps it doesn't matter; I suspect they are the same thing.

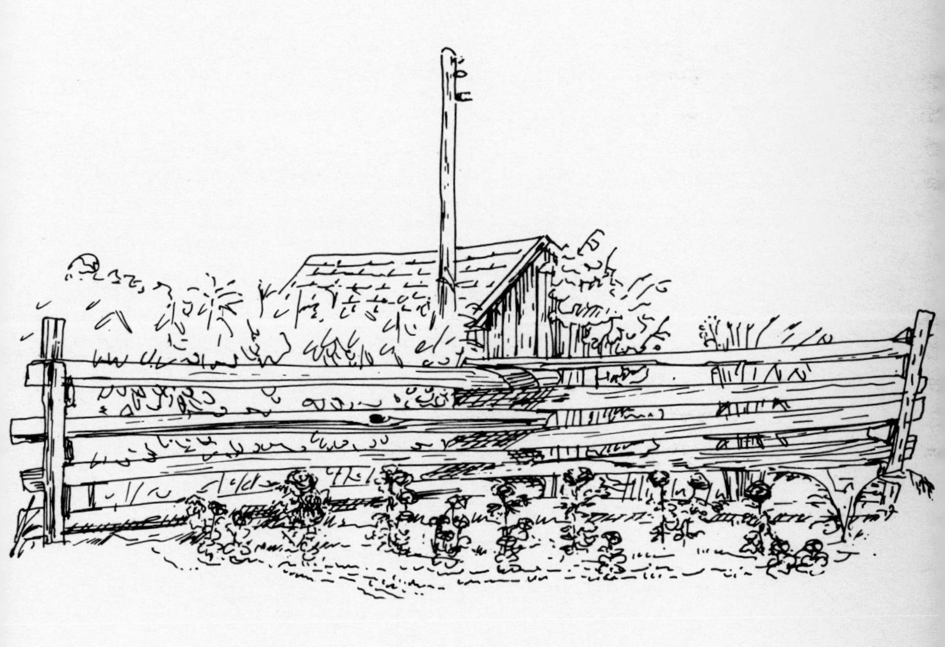

6 HOUSEWORK

There is nothing chillier than an economical and systematic housewife more in love with neatness than with peace. Phyllis McGinley

My mother was so beset with real problems that housework was of necessity trivia. Thus I come by my nonchalance in the face of it quite naturally. An ordered organism, however, yearns for harmony in its environment, and there is always a disparity with which the nonchalant housewife struggles.

Mud and Absolution

I SUSPECT THAT THE KIDS MIGHT DESCRIBE THIS as "the season of the long grouch." Robert Frost referred to it matter-of-factly as "mud time." For him it was April, but every farm wife knows it as an open season from January until such time as the frost is out of the ground and the surface water can drain again to the subsoil. With luck it runs its course in two or three weeks during March, but a January thaw can plague a woman with more than one spell of it.

I have a file of jazzy house plans designed for convenient farm living, and all of them feature a neat little catchall called a mud room. I have such a catchall, but it's called a kitchen! Any day that the temperature rises enough to liquefy the surface, I'm watchdogging at the back door with a broom and a rag and a mean look.

Anyone who has forgotten mud time probably never lived in the country. At my grandfather's farm there were wide boards stretched from the house to the barn to the granary to the chicken coops to the milk house—everywhere—to keep you from sinking in mud to the knees. You left your galoshes in the woodshed or on the back porch, but Gram still railed about the mud season.

Now, of course, sidewalks cut down the tracking on many farms, but they never caught on here, and what little boy is it who can discipline himself to a sidewalk? If you blacktopped everything in sight, the children would haul up a wagonload from the neighbors and dump it outside the back door just to show you what wonderful stuff mud is.

It is unfortunate that mud time and kite time are for the most part simultaneous. Was there ever a kite that flew on the first trial without at least one or two muddy treks back to the rag drawer for more tail?

Like flies to the milk house small boys are drawn to sloppy, slippery, gooshy mud. The higher the boots, the deeper the mud. To "peel out" on a bicycle in a wash of mud and send a spray backward over an admiring gang of buddies is just about the coolest thing a kid can do for fun on a March day. Or to take a long pole and vault across the ditches in the pasture, and maybe not make it—could anything be more challenging at age nine? To wade gingerly through the barnyard in your first pair of rubber boots and feel them sucked down as you try to lift your feet—gosh, that must be a pleasure a city kid would trade you thirty miles of sidewalk for.

Why is it, then, that mothers have to spoil it all by infernally screeching, "Get out of my kitchen with those muddy feet!" Mud! It seeps into the soul as it dries and filters through the rug at the back door.

But now the birds twitter of a morning carrying the secret from tree to tree, to the puddle by the well, to the telephone pole, to the eaves of the barn, and to the horizon beyond. It's a secret they share with these little "mud runners." And when Dane calls me to "come quick!" if

I want to see a pair of robins, I realize with happy resignation that mud is only the penance one serves for the absolution of spring!

Moths and Eagles

One of my spring rites is rereading parts of *The Wind in the Willows,* Kenneth Graham's splendid book about four altogether "human" little animals. In the opening chapter we encounter Mole laboring at his spring cleaning; he feels the "Spring moving in the air above and in the earth below and around him, penetrating even his dark and lowly house with its spirit of divine discontent and longing." The feeling is as familiar to me as the freckles on my hands. Long before I read this perfect expression of it, I had often been tempted to do as Mole subsequently does—to fling down my brush with a "Hang spring cleaning!" and go off to answer that imperious call of "something" beyond the confinement of the house.

Never was I more mindful of what one sacrifices to spring cleaning than the May morning some years back when a nature-loving friend called to ask if I had seen eagles flying in my corner of the wildwood. I observed sadly that one seldom sees eagles flying who is totally dedicated to a search for moths.

Mole took himself away to the river that he had never seen. There he met Water Rat, who became his bosom friend, and life was never the same again for either of them. And since the day when I was challenged to abandon moths for "eagles," my life has changed too. I have relegated the "spring" cleaning to the months of late fall when being confined away from wind and weather belongs to pleasure, when months of outdoor neglect have rendered closets really worth cleaning.

On those spring days when I go out the back door

and find myself loath to return, I redirect myself to the "eagle" search. Sometimes I go like Mole to watch the swollen river; sometimes I find a sheltered spot below the brow of the hill where the grass is already thick and I can soak up the early sun. Sometimes I abandon an afternoon to planting daffodils in the woods (it's not the right season, but they grow). I may spend half an hour "worshipping" my beech trees. Or perhaps I'll combine one or another of these pleasures with a survey of the wild flowers. But never do I forget that I am dedicated to a search for eagles, and I watch continually.

I have come to know the turkey buzzards and marvel at their rhythmic ballet of death. I have wondered after the chicken hawk watching high and alone from a dead tree: "Could that be an eagle?" I have even questioned the blue herons with their necks leading, their long legs trailing, and the unswerving geese, ever fearful that an eagle would fly and I would not recognize it.

"Never mind," said Paul. "When you see an eagle you'll know it!"

At last I have seen my eagle, and he was right. I came upon him one day while driving along a country road. He stood about fifteen feet from the berm—great white head, white tail, and much browner than I had expected—a wondrous big creature.

"Look! An eagle!" I cried to the children who were with me. Then with a flap of those monstrous wings he took off, gliding like a small Piper Cub and coming to rest in a tree on the edge of the woods beyond the meadow where first I spied him. There he sat with wings folded at his side, a massive and imposing presence even at a thousand yards. I watched and watched, my vision blurring, and still I was reluctant to go on while he sat there.

"So it's an eagle. C'mon Mom, let's go."

To me it was an EAGLE, but to the children it was just an eagle. After all, they haven't waited half a lifetime to see one; but then they've never spent any time looking for moths either. We drove on.

Men!

In the way a housewife has of budgeting her time, I allot a few minutes a week to self-pity—the time it takes to turn all the socks right side out. I always suppose, perhaps wrongly, that this is a curse peculiar to a household of men. Girls, I tell myself, learn early the sexy art of running a finger down their heels and languorously removing the sock by its toe (and if they don't they should!), leaving it right side out.

Anyway, this is the time I reserve for bemoaning my fate as the slave to a bunch of men. There are four haphazard laundry piles, each about two feet high. My laundry I can carry upstairs between my thumb and forefinger.

There are eighteen pairs of jeans and three pairs of underwear. What manner of slobs are these? When the boys were infants I changed their diapers as frequently as any well-bred mother should. And during the formative years I lectured on cleanliness and Godliness and the importance of changing underwear. Then why this small sad gray heap? Men!

The socks themselves are a source of trauma—forty-seven socks, one each from about forty-three pairs in varying shades of dark. I often wonder if anyone would notice or even care if I didn't turn them. By the law of averages they'd be right side out fifty percent of the time.

But no need to dwell only on the laundry with its incidental yield of nuts and bolts and coins and jackknives (which chip great hunks of porcelain off the tub). Upstairs on the sewing machine sits a pair of dirty workshoes. Why? It was an empty surface and handy, that's all. There are some ragged tennis shoes on the dishwasher along with eight old spark plugs. Old? Well, that's the problem with spark plugs. Mothers can't tell whether they're old or new. It's the same thing with the flashlight batteries men scatter about counters and dresser tops in great profusion. You can't tell if they're coming or going.

Many of the plagues of an all-male household concentrate in the cellar. There is the indelible memory of coming down to find the washer housing a batch of bullfrogs—or is it a herd of bullfrogs?—anyway, there were a lot of them.

The enthusiasm between age eight and fifteen is bicycles. If you dismantle a bicycle in the cellar nobody will bother it for weeks. Good old Mom will just hang the sheets around it. And she's much more reliable than anyone else at telling you precisely what happened to the hick-a-ma-jig that fits on the thing-a-ma-bob that you left on the third step from the bottom of the stairs.

A ring in the tub, unmade beds, and unhung clothes are standard in a male household. On the rare occasions when I'm fed up to here about such things and complain, my husband says what he's said for fifteen years: "It's your own darn fault! Don't make their beds! Let them make their own. Let them hang up their own clothes."

So I don't make their beds or hang their clothes. And neither do they. Who cares about unmade beds and messy clothes? There's a world out there and life to live, and who wants to surrender any of it to such trivia?

Mother cares, and mother does, and mother makes the beds and hangs up the clothes and collects the nails and screws and empty flashlight cases. She empties the gym bags of wet and reeking gym clothes. She cleans the ring from the bathtub and disposes of the gasoline left in the drinking glass. She empties the jars full of forgotten lightning bugs and disentangles the fish hooks from the bedroom curtains.

She turns the socks, empties the sand from the toes, and on her better days ponders that it could have been worse. She could have had daughters who litter the house with curlers and makeup and twice as many unhung clothes, who also have distaste for making beds and hang on the phone by the hour.

At least around here you can always get into the bathroom even though the place is littered with dirty towels and the john seat is always up.

"The Wash That Was"

ONCE UPON A WASHDAY . . . MY CLOTHES hung on the line at 9 A.M.—all of them. There were no stray pieces left for next week to be slept on meanwhile by the cat. There were no lone socks begging for mates, which would subsequently be discovered in the bottom of the boot box, or under the lilac bushes in the spring. Neither were there any socks turned inside out, "washed only on the inside" (as Orrin has observed) to sprinkle sand and gravel on the clean clothes in turning and sorting. No shoelaces were twisted about the straps of the undershirts and intertwined with the apron strings. No bleach was dribbled across the overalls. Nobody left a book of matches in his coveralls, or a Kleenex in a pajama pocket to disseminate through all the colored clothes.

Every buttonhole had a matching button. There were no irksome decisions—shall I send it around again, throw it away, or put it in the mending until he outgrows it and *then* throw it away? No dollar bills floated to the top of the washer, and no coins sunk in its depths. Nor did I find on the bottom any nails or screws, pebbles or pocket watches.

There was no grease, no manure, no tar, no grass stain. There were, in fact, no overalls, no football pants, no diapers, no white shirts. There were no little boys to climb up by the wringer and ask, "Whatcha doin'." There was not even a husband. There was just a coed a thousand miles from home doing her first machine washing. What could go wrong with a coed's wash? Three pairs of white pants, three pairs of white socks, two white blouses, two white bras, two white slips—and one *red* petticoat to turn the whole mess pink! *Miséricorde et sacrebleu!* (It was a French school.) What did I do? I wept a little, giggled a little, and wrote home to Mama that nobody, but *nobody*, ever had problems like mine!

Rhapsody to a Clothesline

I SUPPOSE I SHOULD BE ASHAMED TO TAKE THE coveralls from the clothesline on the third day's hanging and find a bird's nest abuilding there; but I am only sorry for this ambitious wren who has dissipated her efforts. Beyond that I rejoice anew in the pleasures of having a clothesline. I have heard it said that in some neighborhoods the clothesline is zoned out. Deliver me from such an environment. A dryer is a wondrous machine, but it offers only utility in exchange for a world of aesthetics.

My clothesline commands a view of a hillside pasture sloping to meet the woods climbing the valley. Where they meet in a decadent rail fence, a herd of cattle shade themselves and scratch their necks under shagbark hickory. When I was a bride newly awed with the pleasure of hanging clothes with such a backdrop, my husband and my father-in-law one day moved a chicken coop in line with the view. I protested and they laughed until Aunt Clara Portman happened by to find me weeping in the backyard. She marched upon them with the commanding presence of a Charles de Gaulle and they moved it in shocked submission. How could I ever fail to appreciate this vista?

I could write sonnets about the joys of hanging pastel and print linens against a blue sky next to a cherry tree in bloom. I have been so impressed on occasion that I have nearly stood on my head in a washbasket to photograph the scene. An artist friend once expressed a fondness for painting clotheslines, and for a time thereafter I took particular care to arrange things with an eye to artistic appeal. Busier times, more children and clothes, played havoc with that dalliance. Some days I have counted myself lucky to have time to toss the last washer loads helter-skelter over the lines.

A clothesline is a gateway to a garden. Each flower in its season is accorded due attention because domestic discipline brings the housewife here. I often find myself involved with wheelbarrow and hoe when what I had in mind was soap and starch. Between the syringa and the lilac a spider often spins a fabulous web that glistens in the morning dew, and with the order of his life shames the hapless heap of mine.

I have hung clothes on the line in the middle of a summer night and known there beneath the stars the depths of absolute solitude. I have hung them in the dusk and retired to the house content that darkness saw the day's labor accomplished, then been dismayed when a gale of wind erupted in a torrent of rain. I have seen in lightning flashes handkerchiefs and undershorts sailing over the valley, my sheets atop the chicken coop, and gone to bed resigned to a clothesline crisis that would right itself in the morning.

One of the warmest of clothesline joys is the hanging in the sun of old baby clothes freshly laundered in anticipation of a new life soon to be wrapped in them. When in the course of time that new child grasps the trailing corner of a satin-edged blanket hanging from the line and settles down to sleep in the grass against a gentle dog, there is something about a clothesline that dispels the feminine mystique, that fulfills the housewife in one of her most ancient rituals.

Blue Denim Uber Alles

After the manner some sociologists have of dramatizing statistics, it could be reckoned that the average farmer's wife will spend four solid months of her life patching overalls. When I think of all the broken needles, the tangled bobbins, the ill-adjusted tensions, the cryings-to-heaven in a blue rage, the miles of denim worn to carpet rags involved in that staggering total, I can only tremble. When all's said and done, however, they may have proved some of the most rewarding months of my life.

Bring up the subject of mending overalls in a group of country women, and you stir a lively controversy. You learn as I did at a Church World Service mending day that everyone knows the proper method and that each method is different.

As proof of the superiority of *my* method, I jokingly pointed out that I had been stopped on the street and complimented on my patching by a total stranger. I am of the school that advocates patching three quarters of every worn leg with new patching. The "back-of-the-leg" school (use the back of the leg of worn jeans to patch newer ones) took issue with me, despite my insistence that putting on new patching would make it possible to wear out the back of the leg too. Of course, you'll eventually have to patch the seat. . . . And so it went.

Aside from kooky strangers like the one I met on the street, no one really appreciates your patches except you. The kids keep shifting the mended pants to the bottom of the stack (especially if there's a patch on the seat), or they throw them in the wash unworn. Enraged is the husband who jams a foot in his pants and finds a pocket on the inside knee. He'll complain about the added weight of a patch and the excessive warmth of two thicknesses of leg. Your friends all think you penurious and your time ill-spent.

Nonetheless, there's something intensely satisfying about a stack of well-patched overalls. Perhaps it relates to a passion for salvation. Perhaps it's salve for the wounds of waste in so many other areas—waste of food, talent, money, time. . . .

The farmer's wife who abandons her overalls as a waste of time, who takes a job and discovers she can earn enough in the space of a mending to throw away the old for new, is the "smart" woman. I don't know that she's the wise one; if it's true that our fate is inextricably wound with blue denim, I would not be the one to tamper with fate.

Crisis of the Cleanly

THERE'S A GREAT DEAL OF GROUSING NOWADAYS about the bearded, blue-jeaned, barefoot generation of college kids. But anyone who thinks they are also a dirty and unwashed bunch hasn't had his ear to the water pipe. We had ten or fifteen of these bearded lads and barefoot lassies doing archaeological research here on the farm one summer, trying to live under the limitations of a country water system. Nothing panicked them faster than having their well run dry. In three days' time ten of them went through 1000 gallons of water, which by rough rural calculation is about three milk cans each per day. One can only conclude that it takes a lot of water to shampoo a beard and scrub grass stain from bare feet.

One of the things a country kid dreams about (and hastens to indulge on his first visit to the city) is filling a bathtub to the overflow and contemplating himself submerged to the chin, nobody hollering in the background, "You're gonna' run the well dry!"

Going easy on the water is a fact of country life. You learn early that a bathtub of water can easily do for two or three or four, or more, depending on the season and the state of the rainfall. It has been years since I ran across a washpan in a kitchen sink, but for all the years of my childhood that pan provided the daily bath. The bathtub was pretty much a Saturday-night treat.

With a household of nine the well was frequently dry, precautions notwithstanding. We learned much about saving water that I am often tempted to send as advice to people who complain in letters-to-the-editor that they are on the short end of the city water supply. We didn't water the lawn, for one thing. The crabgrass flourished just as well without water and was just as green as regular grass. We didn't wash the car every week. (We didn't, in fact, have a car to wash.) We learned a few things about conserving on laundry. We didn't wear asinine things like white Levi's and sneakers. A bed only got one clean sheet a week, and three or four towels served the whole mob along with the roller towel in the kitchen. We didn't change our socks and underwear every day either. There weren't that many pairs to change into!

By today's standards I guess we were a bunch of slobs. But we never lacked for friends, so we must not have been too offensive.

We learned that you can flush a john with the water baled out of a bathtub, that dishes can be washed in the water drained off the sweet corn, that a teakettle full of water can bathe the whole family if necessary, and that in dire emergencies you can dip water out of the storage tank on the back of the john to brush your teeth.

We learned that under no circumstances do you turn on a faucet and just leave it running. We learned how to turn a pail and drop it down a well for bailing. But the most important thing we learned was that when you have company from the city, you must watch them like hawks because they use water like it's going out of style.

Of course, I'm not living in the Dark Ages either. I've learned that if you save the rinse water from an electric dishwasher, you can use it to scrub the floor, water the flowers, bathe the kids, or wash white Levi's and sneakers!

The Passing of the Pickle

EVERY YEAR IN AUGUST WHEN I GET OUT MY grandmother's pickle recipe and ponder the challenge of making "a brine that will float an egg," I am more convinced that there will never be a renaissance in the inexact science of pickle making.

There are more drawbacks than the interpretation of

these ancient recipes. In the first place, the cook is told to "put down in a crock for two weeks a peck of dill-sized pickles." Shower gifts these days seem to run more to nighties and negligees than to pickle crocks. And "down," of course, implies a cool, dark cellar, which is harder and harder to come by. I also question how many brides today can identify a peck basket and have any idea where to find one full of dill-sized pickles. I wonder, for that matter, how many girls know what pickles are before Aunt Jane or H. J. Heinz perform their alchemy on them.

After pickles have been "down" in brine for two weeks, it takes a pretty strong stomach to face them again. Much as I'd like to forget, I still remember my first batch! The smell and the mold around the top were bad enough, but there was a round hole chewed in the middle of the cloth I'd tied over the crock and I couldn't face the prospect of finding a drowned mouse in that mess. I threw them quickly in the garbage heap and ran.

I've always planned to calculate the volume of a walnut in order to determine the specific amount of alum my grandmother had in mind for these pickles. But how can I be sure that walnuts aren't bigger now than they were in 1900? And she doesn't say whether she figured it with or without the shuck. Walnuts today come all neatly shelled in small plastic packages. Another generation may be completely baffled by "alum the size of a walnut." "What the Sam Hill *is* alum?" I can hear them saying.

Somehow I can't see my daughters-in-law fooling around for five mornings draining syrup off a crock, heating and returning it to the pickles. Perhaps on some future date in a nostalgic moment one of my sons will say to his wife, "Why don't you make some of those sweet pickles like my mother used to make?"

"Sure I will, dear," she'll say, "as soon as I've sewed you into your underwear for the winter."

Evening Star and Mason Jar

It's that hour of the night when anyone still up and working begins to feel sorry for himself; and I am still cooking on all four burners. This is harvest season and the last chance for "grasshoppers" who have sung all summer to get jelly and pickles and fruit into jars and down to the fruit cellar.

As many a mother has discovered before me, canning proceeds faster during those late late hours without routine interruptions and that juvenile chorus of, "Can I help you?" The counters have been cleared now of supper, gym shoes, lunch boxes, noon mail, evening papers, buckeyes, books, and little boys. A new sort of confusion has spread across them—cans and lids, paring knives, tongs, measuring cups, big kettles, a bag of sugar, a pickle crock.

The sweet pickle syrup has been drained and reheated for the third night, and the heavy sweet smell of vinegar and spices mingles with the aroma of grape juice and the acrid smell of ripe tomatoes. Will this fragrance someday be a part of the memory register of those who grew up here, I wonder briefly.

The jelly bag has been cut down from where it hung all day on the washline inviting the slap of every passing child; the glasses clink together in the kettle where they are sterilizing. Time to make the jelly while the tomatoes are processing.

Yes, the tomatoes, scalded and peeled in the few free quarter hours of the day, are at last in the jars. The well-ordered housewives did their tomatoes last month when they were at peak season, juicier and better flavored. But Disorder and I have long ago made peace, and during these midnight hours she whispers happy reminders in my ear. Order, she says, doesn't have time to make cinnamon toast for all the neighborhood. She doesn't have two or three children perched on her counter while she bakes cookies, or trailing along in a pickup truck on produce deliveries. She doesn't have time to sit for a joyous half hour and watch suntanned boys take an unseasonal splash in the river. Order, for that matter, doesn't know the quiet peace of canning at midnight fanned by a south wind and entertained by a chorus of tree toads, crickets, and katydids.

The grape juice bubbles up, and as I stir in the sugar I realize that somewhere between the pickles and the tomatoes I ceased feeling sorry for myself.

Penny Thoughts at the Window

TWILIGHT OF A DAY IN NOVEMBER, WHEN THE time of the day matches the season in mood. From my post at the kitchen sink I am bemused by the changing, changeless scene. I was once quite unhappy with this kitchen window, so high that I couldn't see what was happening down in front of me. But there's something therapeutic in always having to look up and out. When I stand here, birds and jet airplanes are my preoccupations rather than children and backyard clutter.

"Gosh, I never knew there was a field over there beyond the river," I can remember Dane saying one fall when the leaves had disappeared. It was then I realized he was growing up; it was the first year that he could see over the kitchen windowsill. So many things are clearer when we grow up a bit.

The leaves are gone again and, yes, many things are clearer than they were. There's a field over there in which I've never walked, and yet in a sense it belongs to me. My eyes feed on it through late fall and winter and early spring. This year there is a house there that didn't exist when the leaves came in spring, and new neighbors I haven't met.

On our side of the river, at yet a safe distance from society, the tree house whose progress I followed through all of an exciting summer day is bared now to fall. I was finally invited to visit it, and went with great pleasure, for I love climbing trees and sitting high where everything is seen from a new perspective. The tree house was abandoned with the summer dreams, so perhaps I can claim it as I claim the field yonder.

A few tenacious pears keep faith with the naked pear tree—harvest! What a bounteous, burdening miracle it was! A hornet's nest hangs in the sycamore—"doom" that no one suspected. And who could have known that the apple tree housed four families of birds, or that the ginkgo tree was host to an oriole? Gee, why am I always so sure that I know everything?

Stone Walls of Freedom

IT'S A WINTER MORNING AND I AM CLEANING the cellar. My cellar is one of those old stone monstrosities that you never really "clean"; you sweep up the crumbs from the crumbling mortar and in a week or two it all has to be done again. Leaves and dirt lodge in the broad stone outside stairwell built for the easy movement of vinegar barrels, apple and potato crates, and laundry tubs. Cobwebs collect dust among the electrical wiring threaded through the beams overhead. And a large portion of the floor space is taken up by that granddaddy of all dirt harbingers, a furnace room with coalbin. (Some year soon I expect we'll be running field trips of school children through this cellar so they can see first-hand what a coal furnace and a coalbin are!)

Cleaning the furnace room, I am reminded very much of the village women of India who crown their cleaning activity by sweeping smooth the dirt in front of their huts.

My cellar is also a sort of game preserve. Mice and spiders and thousand-leggers are its most common tenants; when a rat moves in we do take steps to eradicate him.

In a damp summer a toad frequently hops about under the water pump. I am almost the only one who ever sees the snakes. They move very fast when a female screams, but I have one gray hair for each one I have encountered sunning himself on the broad stone windowsills or intertwined in the rafters overhead. (Paul says they keep the mice down.)

It's a pretty gruesome place, all in all. The only bright spots in it are the white appliances, which seem somehow anachronistic—the washer, the freezer, the dryer, the water heater.

I used to rant about the place when I was a bride, and for my nagging succeeded in getting a cement floor and a wall around the furnace and coalbin. But I don't nag anymore. Contentment lies in learning to accept what you cannot change. Now in the more mature eye of middle age, my cellar seems right for a farmhouse.

When a mother goes to the back door and confronts a mess, where does she send it if she dwells in the tidy splendor of split level? Can you skin a muskrat in a basement recreation room? Would you hang geraniums or store gladiolus bulbs there? Where do you go to crack walnuts? Where do kids build their first hot rods? To what unsacred place do you banish them when they attempt to paint a birdhouse on the living room carpet?

Where can a gang of kids come in out of a blizzard and leave a few puddles unnoticed? Or where do you send them to disrobe when they're up to their crotches in spring mud? Where can you make a place for a bucket of bullfrogs? On long winter afternoons little children need a place to ride tricycles and roller skate that is not out of the mainstream of their mother's activity.

A home needs a place that is not sacrosanct, so I no longer resent this old cellar. It is in a sense the "black soul" of the house. And like all souls, from time to time —but not too often—it needs to be put in order. What a satisfying task that is for a winter morning.

At Odds on Order

On some few splendid occasions throughout the year a housewife is overwhelmed with the pleasure of having everything in order. Order is a place for everything and everything in its place; order is clutter set at right angles. Order is the clothes washed and sorted, ironed and mended, and back in the drawers. Order is a clean rug by the kitchen door, no ring in the bathtub, no papers in the living room. Order is a meal planned for tomorrow; order is pink blossoms in a pewter vase on a square of clean white linen. Order is a longing in the soul, an ordered organism yearning for harmony in its environment. And when it comes, it is a satisfaction that envelops one in peace.

Anyone can live in order—for fifteen or twenty minutes. Then the back door slams and a pink petal falls on the square of linen. If you are a country wife, it is probable that standing on your clean rug, dripping wet, are two boys who have been crawling through a soggy pasture after a baby rabbit (and "Guess what we have in this box!").

Already there are two sets of dirty clothes, presently there will be a ring in the tub, and eventually two children and a rabbit will turn the whole place into turmoil. On such occasions the shouts you raise are in direct contrast to the peace you felt ten minutes earlier. You suddenly hate yourself for making a fetish of order.

But it doesn't take a bunch of kids, a slouch of a husband, or a careless housewife to play havoc with order. Even one small dog can do it. Walk to the mailbox and you discover that the dandelions are taking over the lawn. (God and man are at odds on order.) Go out for groceries and you may notice that the car needs cleaning. Perhaps you'll observe that your neighbors are turning your neighborhood into a slum. Answer your telephone and you'll find that someone else has plans for your time that shatter your small realm of order.

If you are the sort of super being who has his immediate environment perfectly controlled, you'll probably find yourself disturbed with the rest of the world, for it's out of step with you.

Anyone can live with order, and if you learn to cherish the moments of the pink blossoms in the pewter vase on the white linen, you can carry a sense of order in your heart. But to live with disorder, to wade through and progress and make a contribution to life in spite of it—this is challenge, and this is what life really asks.

7 THE WORLD

Any man's death diminishes me, because I am involved in mankind . . . John Donne, DEVOTIONS XVII

No one deserves the privilege of living on a remote pastoral "island" blithely unaware and uncommitted to the world's needs. Involvement for the Leimbachs is as simple as a contact with neighbors at the Little League game, as complex as a friendship in the black ghetto. It is as near as the Brownhelm Church or as far as the vision of that church . . . and as challenging as the knowledge that we help to formulate the vision.

The Country Church

IF YOU READ THE DENOMINATIONAL REPORTS you get the impression that the country church has fallen on hard times, that it is withering on the vine and about to pass. But looking down from the balcony of our small white frame church on Sunday morning, I have hopeful notions on the subject.

There is an indescribable feeling of belonging to these people, an awareness that what you have and what you are is largely what they have given you through what they are and are not.

I didn't grow up in this church as my husband did, so a part of its tradition is not mine. But here I grew to maturity. Here Paul and I were married before an altar flanked with evergreen. Here we brought our children and dedicated their lives to God. The people stood and promised to sustain us in their training, a mutual pledge to which we've been faithful. I ask no greater benediction to my life than a memorial service in this small church on a country afternoon with those who have loved me sitting in tranquil silence.

We have worked and played and laughed and cried in this building, sung and taught and argued and prayed with these people—no one surrenders all that without a fight. There's a tenacity in the country church that will not yield to a gloomy statistic.

These are my people and we have no illusions about each other. If I come in knee socks and a ten-year-old coat, they will accept me. If I am splendid in new finery they will rejoice for me without envy. The country church lacks pretensions, or understands well the few it may have. It doesn't attract the pretentious, and that may account for its strength.

A baby cries in a country church, for there is no crying or crib room to which to take him. Perhaps this breaks the "worship mood"; but there is more worship in the united concern of the congregation for the baby and his parents than many services attain in elevated rhetoric. I wonder if one can truly pray in a sanctuary where a baby cannot cry.

We fret because the program isn't what it should be, the facilities inadequate. We think with longing about opportunities offered in big-city churches, overlooking the essential truth—that this is a group where people matter to one another. If you don't make the effort to roll out of bed on Sunday morning you are missed. The old dogmatic fervor may be gone from our country church, and surely we lost something in its passing. But the fear and the superstition and the casting of judgments which belonged to that "old-time religion" I can live without. I would rather be missed in church on Sunday for my mortal person, sinful and human, than for my immortal soul, sanctified and pure.

I read in the church history of 1919 that "Six or more years ago an alarm was sounded by Press and Pulpit. The country churches are slowly dying." The cry now is that there are no ministers to serve the small and remote churches. That may not be all bad.

The early churches had no ministers and survived on their own steam. Through most of the local church history the congregation looked forward to the time when they might have a full-time minister, much as the early church looked forward to the Second Coming, and they were sustained in the longing.

The small church sends people into the world with the security of having been well loved by a multitude beyond their family, and that may be its most important function in an increasingly impersonal age. God is Love, and for many He is more easily found in a country church with a leaky roof where a bird flies through an open door midsermon, than in a cathedral where a paid choir raises mighty anthems over a bejeweled, faceless congregation seated by ushers in morning coats.

Chicken, Church and Cherry Pie

"ARE YOU MAKIN' THAT PIE FOR US, MA, OR are you taking it down to the church again?" It's a wistful question heard with minor variations in nearly every home with a civic-minded cook, and it has a number of implications. At our house it means that Mom will go off to fry chicken at dawn with her grandmother's iron skillet hanging heavily at her side, and return at dusk with her tail dragging. Perhaps she'll be toting a tin of leftover mashed potatoes or a few tired pieces of someone else's pie. If she gets home in between times it's to whip up a spartan casserole for the family, or to change into a clean blouse.

The church supper is an institution deeply entrenched in rural America. Over thirty years ago my grandmother devoted a couple of days a week to preparing food for camp meetings held by her church. There were the inevitable chickens, which she slaughtered, dressed, and stewed. There were great quantities of homemade noodles rolled out like chamois skins, dried on newspapers, rolled up and cut. And always there were the pies.

Grampa grumbled a lot about the mountains of food that went out of the house, couldn't see why the city folk didn't pull their own weight, why it all had to be carted in by the country people. Clearly he wasn't as dedicated to the "cause" as Gram was; and the "cause" was often obscure to those outside the inner circle.

But the dedicated women who labored over their wood-burning stoves did so in the name of the Lord. If chicken and noodles and cherry pie can get you into heaven, then my grandmother is enthroned in glory.

In quantities paltry by comparison, I still tote food to the church suppers and put in the many hours that such projects demand. But I do not delude myself about causes. This is an era that has redefined Christian witness. Few church members today would fail to differentiate between working for the building fund and working for the Lord. Yet the sort of fellowship engendered by a church supper is vital to the sustenance of community.

A church that has lost its communal spirit has probably lost also its sense of mission and Christian witness. Many a country church that merits the distinction of the cross on its steeple is heavily mortgaged to Chanticleer whose soul through the decades has hovered somewhere over the building.

Post Mortem on Annual Meetings

I HAVE HERE ONE AND A HALF SWEATER SLEEVES testifying that this is the annual meeting season. An annual meeting is a sort of inventory of the organizational year, and the lucky members are the ones who knit!

Various enticements ranging from door prizes to dinners are offered for attendance at these affairs, but not even topless tellers could obscure the fact they are "dullsville."

The excitement has all taken place previously at the board meeting where the dirty laundry was aired, the executive director blew his cool, and ladies and gentlemen were reduced to egos in conflict. I think that playing a tape of the board meeting would enliven the annual meeting of the membership considerably. It would also shed some light on the "proposed budget," a document as devious as the Yalta Pact. I have been a party to financial wranglings, and I am sympathetic to the couching of some items in obscure sections of the budget. This is always done with a scrupulous attention to honesty and a clever understanding of the membership.

The annual report has usually been mailed for your study prior to the meeting, and it is physical testimony to the dullness of the whole. Anyone who has ever prepared a section of an annual report is aware that the nitty-gritty seldom requires more than a few lines, but the literary hemming and hawing stretches on interminably. The absolutely dead "end" is to go to an annual meeting where it is suggested that the annual report be read aloud!

Another proposal of which I am very wary is that in which "we go on record as stating . . ." something or other. These are usually plaudits to people whose reward was in their accomplishment, or they are a pussyfooting way of upholding some controversial principle. They would be more worthwhile stated, "I move that we call the newspaper and advertise that we believe. . . ." Printed "records" must be bulging with good intentions that never saw the light of day.

There are some other standard procedures at annual meetings that do not exactly inspire me. The chairman, who owes the agenda and the annual report and fifteen files of supporting information to the executive director and the paid staff, will offer copious praise for his "fine work this year." (This often differs from what was said at the board meeting!) There will be a certain amount of carping about whatever parent group or upper echelon wields authority over the group assembled. And there will be some castigation of the membership (especially the absent majority) for general lassitude.

The organizations in which I have gained my experience of these matters have always been benevolent in purpose, and it impresses me that somewhere within the business routine of the annual meeting we manage to lose sight of our benevolent aims.

I would make two major recommendations for annual meetings were I to be consulted in such matters. (Both would probably be judged impractical.) First, I would proclaim that the aim of the group be read aloud at fifteen-minute intervals. Second, I would reverse the order of the agenda. The first three quarters of the business is usually devoted to self-concern; if there is any money, time, or patience remaining, we get around to our limited concern for others.

If meetings are scheduled judiciously they can be rapidly dispatched. For example, scheduling an annual meeting on the afternoon of the Super Bowl game is sure to cut debating to about half time!

Richard Armour said it all:

Which motions have the most success?
For which do persons yearn?
One is the motion to recess;
*The other, to adjourn.**

Betty Leimbach

"LET'S GET BUSY AND STUDY MY SPELLING," said Teddy tonight as I picked myself up off the floor.

"Whatta ya' mean, 'study your spelling'? Since when do *we* study your spelling!" I shrieked. "Isn't it bad enough that they're fighting in Israel and Cambodia and Outer Mongolia without our starting a war right here in the kitchen? Why don't you just run along and watch some violent TV show and get the whole business out of your system?"

"Do you want everybody in the fifth grade to hate me?" asked Teddy quietly.

---

* Richard Armour, in the *Wall Street Journal*

"Since when's it been so cool to get good grades in spelling?" I asked.

"Since Betty's offering us all a bottle of pop if everybody gets 100," he said.

"Betty" is the intrepid heroine who teaches fifth grade, and woe be unto the kid who forgets himself and calls her Mrs. Leimbach. And how do you get away with that sort of familiarity without losing your dignity? When you're the best darned kicker on the fifth-grade football team you achieve a status that surpasses dignity, and that is Betty's distinction.

She is the sort of natural teaching genius who injects so much wonder, excitement, and genuine enthusiasm into learning that going home is a letdown, vacations are a bore.

What Teddy loves most about fifth grade, he claims, is working with the jigsaw, the drill, and the sander. Betty found the old equipment rusting in a corner somewhere and coerced the principal into setting it up in her room. It's an even greater inducement to excellence than free pop for spelling.

One day I was coaxed into trucking a couple of empty mattress boxes to school to build a ship for Magellan, the star of an original drama written by a fifth-grade committee. I was inclined to skepticism about the educational value of the project, so tonight I queried Magellan on explorers. He rattled them off as though he were a pre-Renaissance scholar instead of a frustrated jigsaw operator.

Teddy's fifth-grade room hums constantly with projects, perhaps an electric skillet rigged for an incubator or a mechanical Rudolph wired with a flashing red nose; off on the stage somewhere a group works on a puppet show, cautiously disciplining themselves because that's part of the grade and "that's what Betty expects."

At lunchtime when most teachers delight in a few minutes of well-deserved relaxation, Betty goes to the playground or the gym with her brood, or takes a walk alone around the schoolyard to burn up a little of her pent-up energy.

Before the year ends, Betty will organize a basketball tournament for all five sections of the fifth grade, and every one of the 150 or so boys and girls will play—the lame, the halt, and the blind included. And maybe someday she'll get them all down in the gym and teach them to dance, because Betty Leimbach thinks that in living a rich, full life dancing is as important as arithmetic—maybe more so.

A few days a year Teddy will come home and say, "We had a substitute today. Betty had to stay home and plant wheat or pick corn or plow." Before Betty was a schoolteacher she was a farmer, and farming remains as important to her as breathing. "Chores come first," she tells the children, and whether or not one agrees, one understands that she does not imply the neglect of schoolwork.

Or perhaps she will have stayed home to butcher. It takes a lot of meat and canned foods to feed a family ranging between six and twelve (all male except Betty), depending on how many are home from college or the service.

And in between, during, or despite all the good times, Betty's fifth-graders become respectable little students. Teddy was called to the phone one night to relay some vocabulary words to another fifth-grader. We all sat with our jaws dropping as he pronounced them: ". . . obstreperous, herbivorous, carnivorous, aquatic, loquacious, laconic . . ."

"Obstreperous! Herbivorous! Loquacious! Where in Sam Hill is she finding those words?" asked Paul.

"I'll tell you where she's getting them," said Dane, flipping open his senior English notebook. "Right there. She's getting them off of Jack's senior word list." (One of Betty's nine sons is in Dane's English class.)

"Do you know all those words?" Paul asked as Ted rejoined us at the table.

"Of course I know them," said Teddy. "When Betty grabs you by the hair and says, 'Teddy Leimbach, are you working as hard as you can work?' and you hafta' say 'naw,' well you know you better get busy." And then he proceeded to define those words. I didn't hear all of them; I was quietly fumbling through my dictionary to find out what "laconic" means!

Betty is a corn-fed dynamo with more energy than six average women. The key to her success as a mother, a teacher, a farmer, a friend is that she believes in getting right in there and learning or playing or working or "having a ball" with the crowd. But if I were to give a "laconic" description of Teddy's fifth-grade teacher, I'd have to say she's "the neatest, the coolest, absolutely the most." My best days are those on which somebody mistakes me for Betty Leimbach, to whom I'm related only by a valued friendship.

En Garde!

DURING THE WAR YEARS AND LONG BEYOND into the Joe McCarthy era, we were cautioned against a subversive element within the nation, a fifth column of sorts. In recent years we have been so preoccupied with demonstrators and hippies that we are dangerously lax.

At the risk of starting a new witch-hunt, I feel compelled to call to your attention an element as insidious as any ever dispatched from afar: "they," the anonymous "they"!

"They" are everywhere, running everything, and making a dreadful botch of it! "They" run the churches, forever increasing the budget, making ridiculous appropriations and irresponsible decisions. "They" also run the PTA; "they" have made of it a meaningless farce. (And speaking of misappropriations, well, I tell you. . . . "They" always seem to make their biggest boo-boos in dealing with "our" money!)

"They" run the schools (and do it badly!), "they" have a hand in running the hospital boards (and levy those sky-rocketing rates). "They" are the parents of the delinquent children who tend to corrupt "our" wholesome ones. "They" are, in fact, that group of "everyone else's" parents who never set limits on where their kids go or what time they come home.

"We" won the last war, but "they" are losing this one. "They" have made such a mess of things in government that no one will ever restore order. "They" create waste and are to blame for pollution.

"Their" morals have declined so sadly; "they" read all the "dirty" books, "they" permit and encourage the lurid, the obscene, the violent in every media. "They" cheat on "their" income tax and abuse "their" insurance policies. "They" are constantly in demand of higher wages and shorter hours. "They" make impossible and increasing demands on the government.

A subversive element of incredible proportions indeed! And yet try to identify "them."

I, of course, have never been a party to error, and certainly not to any careless disposition of funds. Everything I put my hand to succeeds. My principles are the highest. If I have any vice at all it is only lying.

And you, of course—none of this depravity derives from *you*. Who are "they"? Something should be done.

I have thought and thought about it, and the only solution I discern is that "we" who are omniscient and omnicompetent, whose morals and principles are above reproach, must get in there and take charge!

Palsy-Walsy

As THE URBAN CULTURE CREEPS UPON THE rural community, it is inevitable that one day a farmer's son comes home and announces that he's joining the Cub Scouts. Now, no good farmer with fifteen or twenty years of 4-H to his credit is going to take kindly to this turn of events. Just as he has reservations about urban 4-H clubs that specialize in dogs and riding horses, so he frowns on his farm kid's meddling in a pursuit that is traditionally for city kids—namely, Scouting.

But a nine-year-old has a way with his mother, and mothers have ways with dads, and sooner or later they're all at the first pack meeting taking the pledge. The 4-H'er pledges his "Head . . . Heart . . . Hand . . . and Health"; but the Cub Scout pledges his dad, mom, brother, sister, and all their resources (which may explain why the urban community is swallowing the rural!).

Part of the theory behind Scouting is that a dad ought to be a pal to a boy during the fleeting years of early youth. I get the distinct impression that what my spouse wants to be to his sons most of all is a father. He wants to come in from chores, take off his boots and coveralls, send his son to the mailbox for the newspaper, and put his feet up to read it while the kids play together on the floor. Mom's suggestions that he should be downstairs building a rocket with his Cub Scout go over like a lead balloon.

It's not as though he were on the road five days a week and abandoned this boy to his mother. After all, the boy has tailed him around the farm like a shadow for nine years. He has demonstrated for his son the dignity of hard work; he has taught him his pride in farming, and transferred to him the love of the land to which he's heir. He has taught his son to fish and swim and ice skate and ski. He has given the boy his quickness of movement, his stubbornness of nature, his lean look, his very name.

Now suddenly there is in the kitchen a shrew who's telling him the kid's going to be a "ruin" because they're not sitting on the cellar steps whittling and getting to be "pals." His dad and he were never "pals" and they grew in time to be good friends. Where do women get these inane ideas?

At the second pack meeting, of course, where they sit with twenty-five other disillusioned mothers and little boys watching a film strip of a boy growing to maturity and responsibility at the side of his father, as they build an ice boat in the garage. And where's "ol' Dad"? Why, he's home being "palsy" with his older son as they finish the chores in the barn.

Baseball Then and Now

HAVING SONS OF NINE AND ELEVEN, I HAVE BEEN liberally exposed to the Little League baseball scene. And I am prepared to testify that baseball is not what baseball was.

We came up through the real "sandlot" days; that is to say that somebody in a neighborhood (for whom kids were more important than grass) had a yard where the grass was worn thin and the sand showed through in great irregular splotches. There wasn't any elevated pitcher's mound with a rubber mat, and the bases were likewise unmarked.

There was, of course, no umpire to settle disputes. Even if there had been enough players to field two full teams (which probably only happened on the occasion of family reunions), nobody would have dreamed of surrendering judgments to anyone less than God or the junior-high gym teacher. Much of the satisfaction of the game derived from those wonderful arguments about whether one was safe or not. Without any bags to step upon, who could say for sure?

The guy who could shout the loudest usually won the point. Not satisfied with that, you could throw down the ball and go home. If you left your team shorthanded enough, you had the satisfaction of knowing you'd destroyed the game; or, better still, they'd beg you to stay. And if they won you over with their, "Aw, c'mon . . ." you were something of a second-rate hero.

A gang collected spontaneously of a spring or summer evening, and a couple of recognized champions tossed a bat back and forth, stacking their alternate fists up its length to determine first choice of players, or "ups." There were certain unwritten rules we followed. For example, if you owned the ball, you got to be the pitcher. If the bat was yours you were first up. (And knowing that you could take either your bat or your ball and go home *really* gave you a sense of power!) The kid lucky enough to own a glove either played first base or was conned into loaning it to somebody who could play the position better than he.

The weakest and the smallest kid was usually the catcher, and he played way back behind the batter. Catchers were considered expendable; in fact, the position was usually filled ineptly by some member of the team at bat. If you were a girl or otherwise unremarkable, you played out in the field. When you had a rare opportunity to holler, "I've got it!" you invariably collided with a second baseman who was convinced that he could do a better job of catching a fly. And who was there who didn't agree with him? (If you think this sounds like the belated grumbling of a left fielder, you're right.)

Those games were never won or lost conclusively. They were usually called on account of darkness, and the apparent winners went home buddy-buddy, talking about getting together tomorrow to form a club or perform some other exclusive rite. The losers quarreled among themselves about who was at fault and agreed only on the point that the opponents were a bunch of "dirty double cheaters." Oh, it was all wonderfully disorganized and satisfying.

And then the adults got into the act and organized the whole business. I think we were always aware subconsciously that the game and the circumstances were imperfect and at fault. Besides, we could always rationalize that we were right, no matter who won the point. Once there was an umpire to declare the right and the wrong, the kids could only direct their spite at one another, at the umpire, or at the coaches. The game improved and human relations deteriorated.

I suppose the thing I resent most is that the spontaneity has gone from the game. I've never been very responsive to strict schedules for myself, so it's no small wonder that I take a dim view of regimenting nine-year-olds (especially when it means I must organize my household around ball games).

However, for all my objections, I see signs of hope. Our scrub team has a 4-0 record (we have the 0) and one might expect feeling to be running high. Yet at last week's game I heard nobody poormouthing the coach. One of the local fathers (ours) was forced into service at the last minute as an ump, and nobody hurled any epithets or bottles at him. None of the kids cried or stamped on their hats.

There was a general sense of pride in these little kids, that they perform with the skill and nonchalance they do.

There was an air of congeniality and conviviality in the crowd, and as the game broke up spectators from both sides laughed and chatted together.

I was impressed that these are parents who still see things in the right proportion and recognize that as an attribute to pass to their children.

Fair Enough

"DO YOU HAVE A GOOD FAIR UP HERE?" ASKED my new friend from Down East.

Gosh, I don't know! Do we have a good fair? I've never been to anybody else's fair; how would I know? It's no Expo '67, but then it's a lot more than a carnival. I've gone there every year of my life. There are years when I was bored stiff and years when I was enthralled. The big difference, I reckon, was not in the changing fairs but in me.

When you are a parent who has sweat blood over a 4-H project along with a procrastinating kid—when you have agonized while she awkwardly sewed up a bean-bag; driven for miles to find somebody to teach you both how to tie a sheepshank; scoured the township in search of the requisite display vegetables that a sad little garden hasn't yielded yet; put in long weary hours helping your child in the care, feeding, training, and hauling of an animal; or stood nervously by while a 4-H judge asked him questions you're not sure *you* could answer—when you've done any of this, then it's a "pretty decent fair."

Should you be young enough to eat cotton candy without feeling conspicuous; to climb to the top of the grandstand and drop water-filled balloons on your friends; to sport a T-shirt saying "Keep America beautiful—wear a litter bag over your head!" and enjoy the attention it brings you . . . or clever enough to have ridden every piece of amusement equipment without losing either your lunch or your pocket change; to muscle into a good standing position along the track and enjoy the grandstand show without benefit of ticket; to connect with the baseball trigger that drops the girl into the water tub; to be a girl holding hands with some boy (any boy) along the midway; to be a boy who wins a four-foot stuffed animal for some girl (some special girl); to take an animal project and spend the whole week at the fair, and go home reeking and grimy and malnourished to smile in your sleep between nice clean sheets—then you're sure to find this a "real cool fair!"

If you have the passion for horses of a thirteen-year-old girl; if you're a farmer thinking about a new tractor and can "flirt" with the machinery merchants more seriously than usual; if your flower arrangement gets a blue ribbon, or your pen of sheep wins a "Best of Show"; if you are the proud owner of the Grand Champion chicken; if your trotter finishes in the money; if you know somebody working an exhibit and he proffers a chair inviting you to chat a spell; if you're pushing water softeners, sell fifteen and get forty leads; if you're a politician with your glad-hand out, and somebody grasps it with sincerity recognizing you as more than just a "politician"; if you're an old-timer in pursuit of an elusive "something" you knew as a kid at the county fair; if you thrill at being surrounded by wholesome and capable young people—then you're certain to judge this a "mighty swell fair!"

If you are none of these people, if none of their feelings are yours, then it's "just another county fair"—buncha' stinkin' animals, lot of cheap sideshows, overpriced amusements operated by greasy-looking characters, fast-talking salesmen out for a buck, too many people, too much noise and dirt, big waste of time and money! Better stay home and watch television.

Bib-Overall Legacy

JUST ASK ANYBODY OVER IN HENRIETTA WHAT HE thinks of first when he thinks of the community and he'll tell you right off, "Harvey Born in his bib overalls and his Cadillac!" Harvey was a great friend of my father-in-law, and we always admired him for being "firstest with the mostest." He lived fully and heartily in the best tradition of the American farmer, but he brought to that tradition his own unique punch (down at the mill or over at the Farm Bureau they would probably have called it "guts"). He early demonstrated that a farmer had better get into farming big or get out altogether. In the late 1940s and early 1950s when Harvey took the plunge necessary to transform an average family farm into a mechanized, computerized dairy business, there were plenty of skeptics around to scoff. A decade later the few who were left in business had been forced to follow his example.

People came from all over the United States and the world to visit the farm at Henrietta Hill and see the incredible innovations that Harvey Born had brought to the dairy business.

I often took foreign friends there that they might understand the force that makes American agriculture great. They came away awed by this friendly, exuberant fellow who took such pride in his place and such pleasure in sharing his knowledge.

Harvey wasn't the sort to put off till tomorrow what he could enjoy today. He and Sarah had traveled all over the United States and Canada on what always turned into busman's holidays. He would drive for thousands of miles to visit a farm operation he thought might give him an idea he could apply back home.

We farmers treasured Harvey Born as a puckish friend and a devoted community leader. But most of all we shall remember him in his bib overalls and his Cadillac as a symbol of courage when we very much needed one. The Cadillac promised us that farmers need not always live as second-class citizens, while the overalls warned us that nothing was going to eliminate work. Harvey himself represented the ingenuity and determination that could bring the first condition out of the second.

Harvey Born improved his world considerably by the labor of his hands, the example of his life, and the family he left to carry on his work. Heaven, too, will be a better place for his being there. You can bet your rubber boots that he has the place fully mortgaged and is busy making improvements.

Any Sin in "Sincere"?

THE WORLD OF BIG DEALS AND WHEELER-dealers is "another country" to me. Nevertheless, a farmer's wife must allot some time to sitting around in offices waiting for the unraveling of the red tape that ties a farm operation to the urban community of business. A farmer's wife who sits and knits in a swivel chair in a factory office soon learns the power of a sit-in. And she may well learn more than that.

It was a bad day at the bag company, no doubt about it. I was the only back-ordered individual bugging them visibly, but there were dozens of others on the phone: "Hello, Harry! How's the wife and kids? Playing any golf? No kidding? Well, what do ya know! Say, Harry, about that order—we're just not going to be able to deliver on schedule. You know how it is, Harry, to get good help these days. Those guys in the back couldn't care less about our customers, Harry. They don't have to talk to you guys. . . . I know, Harry, I know. If it were anybody but you. . . . I tell you, I feel worse than you do, Harry." (Someone in the Dale Carnegie course had told him that the sweetest sound in the language to any man is the sound of his own name.)

And so a couple of hours passed. Each of a dozen calls attempted a sincerity to which the salesman was uncommitted. There's nothing new or foreign about the high-pressured pitch—even a farmer's wife gets a lot of baloney drummed in her ear—but sitting there and listening to the commentary between calls intensified the hypocrisy. Nobody seemed willing to acknowledge blame for anything. Under the pressures of the afternoon, tempers flared and the backbiting increased. Even an incidental phone call to the bookie did nothing to improve the spirits of the office force.

And then there was the plight of the unfortunate field salesman waiting on the hold line to whom the company official was out: "I simply will not talk to that guy! He makes me physically sick—starts ranting and raving and telling me how to run the business. I *know* how to run the business and I'm not taking orders from anyone except the Man Upstairs." (He swept an arm heavenward for the benefit of the farmer's wife on the sit-in.) The question that crossed my mind as I departed with my knitting and my belated merchandise was how he could be so sure that the Man Upstairs wasn't dangling on the hold line. . . .

". . . and He Made Five Talents More."

ONCE UPON A HARD TIME ENRICO GIGLIOTTI and I were students in the same geometry class. Sometimes he helped me with the problems; sometimes it was I who saw through the theorems and corollaries to the logical conclusions. But throughout those study sessions huddled around my dining room table, I detected in him a certain resentment that I did as well as I did with geometry when obviously the practical applications of the stuff were beyond me.

For my part, I was bored with Enrico's probing questions. The only thing that concerned me was that I get an "A," and I couldn't see that it required total involvement.

While I went on to college, Gigs went into the Army and then into the pig business like his father before him. During one especially lean year in pork, he took a two-week job as a construction worker and never went back to pigs. Labor was a subject Gigs knew from the bottom up. At the age of three or four his mother would set him on the drainboard to dry dishes, and at six he was scrubbing the kitchen floor regularly. In that Italian family labor was elevated to an art and considered far more important than education.

On the construction job he so impressed two old journeymen masons that they invited him to become their apprentice; thus did he break into a trade usually perpetuated in those days only by the people born into it. He developed the rare talent (and the rich vocabulary) necessary to get the most work from a crew of men and moved up to labor foreman, later to mason foreman.

Gigs took us out recently to see a school he's building as construction superintendent, a multimillion dollar complex of four buildings under a continuous roof covering two acres—geometry in brick and block and glass and steel.

To someone who can't coordinate a three-course meal, it appears a staggering project. From his trailer office strategically placed for optimum visibility, Gigs pulls the strings that make the project hum. He lays out the job, makes job assignments, orders materials, and coordinates delivery and distribution with the available men and machinery. As we tramped over the building site he explained the project and its evolution.

It's early in the job now, and one day's work may involve excavation, pouring footings, setting precast concrete, taking delivery on a load of steel (and wrangling with a driver about who will unload it), laying brick, and working with electrical and plumbing contractors. Before the students inherit the building, Gigs will be calling in and coordinating the efforts of from twenty to thirty subcontractors.

And all the while he will be the harassed buffer for the architect, the general contractor, the school board, the union representatives, the building authorities, the materials suppliers, the laborers themselves, and all the subcontractors. All will bring him their beefs and look to him for a resolution of difficulties. Even the NAACP may show up to complicate his job. These things I learned as we tramped over the building site and he explained the project and its evolution.

On the way home Gigs and I drove past another school where he had supervised construction, stopping to peer in windows and discuss the work. "For somebody with as little education as I have," he said, "I've spent a hell of a lot of time in school buildings!"

Then the conversation switched to other things—to the stock market, which Enrico knows better than most stockbrokers. I asked him why he didn't go into the investment business, and he shrugged. It was really a rhetorical question, for clearly he had a need to be involved in this manual occupation he handles so well.

You couldn't call Gigs a satisfied man. Society tells him he should have an education, and he is frustrated at that point. I thought about my college degree and how really little it had prepared me for anything; it serves me about as well as all that geometry from twenty-five years ago.

"What," I asked him, "does the architect know about the job that you haven't taught yourself?"

"The stresses and strains," he answered quickly, as though he'd pondered the subject often. Thinking back over the multitudinous personal encounters, the hassles, and the frustrations represented by the two acres of building I had just seen, I smiled. "That," I said, "is a big laugh!"

Anybody for a Riot?

WHEN MY SON TELLS ME THAT HE WON'T GO to school in pants with pegged legs, I give him a tongue-lashing on his ingratitude at the privilege of even owning pants ("Think of all the kids around the world that . . . blah, blah, blah!"). When the smoke clears, he goes out the door in a huff. In the ensuing days I go quietly off and buy a couple new pairs or settle myself to the tedious task of sewing flare gussets into the old ones.

Gregory slouches on a beat-up couch in a stinking hot tenement on a June morning and bugs his mother about the same problem. It is finals week and he is in danger of failing math.

Six little kids wail and cajole for separate attentions in

the steaming upstairs apartment, stifling with the smell of too many people in too few rooms.

It's an asinine business. The pants cover him decently, but he's fourteen and being a black kid in the slums doesn't exclude him from the fetishes of fourteen. He's noticed that nobody who matters to him wears pants with narrow cuffs.

What should this mother say to this kid? "Think of all the kids around the world . . ."? Is there a world beyond East 38th Street?

"I'll run right out and get you some." With six kids hanging onto her skirt? She's miles from a place where she could get a decent buy, miles she seldom has an opportunity to cover, and then only with the six children by public transportation.

"Take $10 and buy a couple pairs yourself"? She tried that once and he came home with a sleazy pair of dress rayons for $6. They tore with the first wearing. What does a young boy know of quality purchases?

These with the narrow legs obviously came from a rummage sale—a good buy, with a brand name label—good quality wool, for a June heat wave.

"Greg," she says, "why do you bug me with this problem when I have so many other problems already?" It's not a question. It's a plea for understanding. "Greg, there's nothing else to wear, and you've got to go to school!" He slumps deeper and stares straight ahead. He's tuning her out. "Greg, what's the matter with you?"

What's the matter? What's the matter? It's a question he's asking about his whole life these days.

And into this impasse between two tragic generations I waltz—the great white mother, with a bag of potatoes under my arm (alms for the poor!).

Greg's mother has some psychic understanding that I've been "sent," and somewhere I must find an answer.

I look through her collection of old clothes for something to use for gussets, then borrowing an iron and a pair of scissors from next door I set about updating the pants.

Damn the fashion industry! Lord, the heat! Most of the windows are painted shut—no breath of air penetrates. It's ten o'clock. What will this be by five?

The small children fret and quarrel and clutch at everything—the thread, the scissors, the razor blade, the ironing cord—cautioned by this interloper who in her brief hour can afford to be patient.

Then it's noon and Greg is on his way to school through the unshaded neighborhood in flares with a jazzy twist. I'm in a hot truck on my way back to the hot countryside, but it's a hell hole I'm fleeing.

There's no Annie Rooney "good all over" feeling. I was an accidental instrument to a single impasse out of a lifetime of impasses. What could I ever be here but a teaspoon against a tide?

I don't know, but I keep asking myself.

Windows on the World

IF YOU KEEP THE FREEZER WELL STOCKED—especially with plenty of homemade bread—you can sneak away from the farm from time to time, even in September. In the fall of 1968 I escaped for three days to Lansing, Michigan, to attend the conference of the Associated Country Women of the World, meeting in America for the first time in thirty-two years.

In the soft and gracious tones of the King's English, the young and beautiful president of the Associated Country Women, Aroti Dutt, told of their work around the world training women as leaders in remote villages, teaching nutrition, mothercraft, homecraft, citizenship, literacy—everywhere promoting international understanding.

These women share a common dream—peace. And looking over that body of several thousand competent, confident women representing 116 countries, states, and principalities, imagining the many thousands whose lives are touched by each of them in real and practical ways, peace seemed much more than a wavering ideal.

A woman from Ireland talked of establishing a school for retarded children. A Canadian shared details of her radio forum for farm women with Indian women seeking ways of serving the illiterate.

A poised and gracious young African from Cameroon (the first woman ever elected to her nation's parliament) told of the problem of combating food taboos based on old wives' tales ("children who eat bananas become infested with worms; women should not eat chicken").

A Fiji princess, regal and stately and sturdy as a silo, related her efforts to teach handcrafts that will yield profit for their tourist industry.

An Indonesian woman described her people as living simple and happy lives on ten dollars a month per family.

A representative from Nepal told of women who walked two days through forests and over mountains to attend clinics on child care and family planning, held sometimes in open fields, sometimes in empty cowsheds.

The story is the same in other countries. In Lesotho a woman thought nothing of walking fifteen miles with a gruel pot on her head and her twin daughters strapped to her back.

Mrs. Dutt spoke of visiting one homemaker's club of forty-five women, all wives of a single chief. Cleopatra Romili from Trinidad expressed the yearning of women everywhere to be recognized and appreciated as individuals.

Sitting there in the sterile high-rise luxury of the Michigan State University campus, we seemed a long way from the frontiers of need we were hearing about. Yet, I reflected, here in the very heart of the technological Mecca, in the core of American cities there was an illiterate, malnourished, deprived populace to be reached perhaps with the same simple person-to-person approach that Sofia Bedwani employs in the villages of Egypt.

Secure in the spacious center of a wealthy, fruitful

nation, Midwestern farm people are not accustomed to thinking in global terms. But we are not strangers to compassion, and the realistic "soup and bread" ministry carried on by the Associated Country Women of the World we understand very well.

One Indian woman made literate through a pilot program of ACCW had said, "It was as though I had spent my whole life in darkness, and suddenly a window had been opened into the world." We Americans are too accustomed to thinking of ourselves as the "light" to recognize our own darkness. Those brief impressive days with the Country Women of the World had opened a window for me too.

"The Steady Life" . . . Alas!

WHEN I MET WERNER STOLL IN 1956 HE was a young man of twenty-nine with a great deal more promise than responsibility. He was part of the executive training program of a small German factory manufacturing motorcycles and bicycles. I sat over coffee and pastry with him and his soft-spoken wife, Trudy, in a remote and lovely inn overlooking the beautiful Neckar River in South Germany. In the patchwork landscape of the valley below us, the twin steeples of a cathedral dominated an ancient village. Black barges glided downstream en route to Heidelberg, Mannheim, and the Rhine.

The scene was tranquility itself, and that is what we talked of—tranquility. Werner was feeling contentment in a world where peace returned slowly and prosperity began to seem possible. He had worked with Americans during the Occupation and knew them well. He chided me about the frantic pace of American life occupied with babysitters and second cars and moonlighting and moving up to split-levels.

And in his limited vocabulary he tried to express the values he and Trudy had established for their lives. Of ultimate concern was "the steady life," a life that allowed time for afternoon "coffee," for long family walks, for sitting quietly in the evening reading to the children or playing the piano for the family to enjoy. "The steady life" provided long, peaceful, August vacations in the Black Forest, walking across fields of wildflowers into the long shadows of the pines, sitting on boulders beside gurgling mountain streams, sipping wine on the balconies of mountain chalets. And in winter "the steady life" prescribed long hours of reading Goethe and a few more days' vacation somewhere in the mountains, skiing a little, sleeping a lot, enjoying pleasant conversation over hot grog.

Ah, "the steady life"! It became a dream to me—a pleasant thought that in itself flooded me with tranquility, and always part of the thought was the peaceful afternoon over coffee, the ancient village, the patchwork landscape, and the river flowing barges to the sea.

Now Werner Stoll is forty-seven. He has fulfilled the promise of twenty-nine and is one of the executives with that same German factory. No longer small, it now manufactures automobiles instead of bicycles and motorcycles, now has merged ("swallowed like a small fish by a whale") into one of the world's largest automobile corporations.

Werner writes of his promotion, welcome just now because it provides funds for a much-needed addition to the house. "Last week," he writes, "I did Stuttgart-London, London-Copenhagen, Copenhagen-Oslo, Oslo-Copenhagen, Copenhagen-Stuttgart." Trudy sits home and "dies" with each flight. Life is a rush of attaché cases, long-distance calls, wrangles with lawyers, high-level decisions. The pride in the small family corporation and his long association with it slowly passes with the identity of that company.

And "the steady life" erodes bit by bit by bit. There is now the problem of the "nervous stomach trouble" and the son who "adapts to some of the foolish habits of teenage." I too am in my forties, fretting over the ways of teenagers. Life is a rush of skillets, washing machines, typewriters, schoolrooms, community responsibilities, and low-level decisions. Someone tells me I look "tired." Worse than any of these single abrasions is the multiple impact of the evils of our age—war, drugs, population explosion, pollution, uncontrolled technology.

I come across these lines in a newsletter: "Slow me down, Lord . . . break the tensions of my nerves and muscles with the soothing music of the singing streams that live in my memory."

Is there really any such thing, I wonder, as "the steady life," except as it exists in the soothing music of the singing streams in our memory. There is still peace in the memory of the ancient village, the patchwork landscape, and the river flowing barges to the sea.

To Know Us Is to Love Us—Perhaps . . .

It is possible to be a guest in a house for a week or two; but when your bed begins to go unmade until afternoon and your clothes are getting lost in the mending, it is safe to assume that you have become part of the family. So it is that after several months at End O' Way, Franz is very much a long-suffering fellow resident. Franz is an ambitious young Austrian whose mission in the United States is comprehension of the English language, and it is our great pleasure to have him as an eager student.

Sharing your home with a stranger from another nation offers hazards as well as privileges (and how can you call anybody strange when he has become as familiar as your kitchen cupboards and as comfortable as an old couch?). First of all you risk losing a large portion of your heart to this individual who gropes so earnestly for understanding and expression in a language not his own.

You are accorded the difficult privilege of interpreting your world in a too-limited vocabulary to someone probably fortified with a bag of misconceptions. You expose yourself to the trauma of seeing your family, your home situation, your nation through very different eyes. In the process you may find that you open yourself to change of habits and attitudes, which finally involves pain.

Americans have a notorious complex about their nation, their values, their standard of living; about the existence side by side here of great affluence and poverty, and about their world image. As though the confession of all our "sins" will somehow bring absolution, we parade them for Franz's consideration.

We talk about willful children and the effect that over-indulgence has on them. We talk of the deteriorating influences of "go-go-go" family life. We talk of the American passion for education and the too-frequent shortcomings of educational aims and processes. We talk of the bourgeois mediocrity of our obsessions with sex and sports and violence, of the peculiar problems of being young in America today. We discuss alienation among generations and races and economic strata. We talk of the bleeding wounds of Vietnam and the American cities.

At the same time we introduce him to the wonderful mélange that is our life in America; to the warm places in community life where we assume responsibility and find acceptance. We bring him our family and our friends one by one and try to express the joy that we find in them, the special contribution that each brings to our lives.

We dissect for him the unique quality of American humor that rescues us so often from the edge of despair. We share our literature as Franz is able to comprehend it, hoping somehow he may assimilate the feelings of a Whitman and a Sandburg in their broad appreciation of our nation in its candor and generosity, strength and vitality, richness and diversity.

We try to explain the forces that have shaped our determination, our prejudices, our present dilemmas. We would communicate our dreams and aspirations—and this is hardest of all across a language barrier—all the while hoping that somehow these understandings, appreciations, hopes, and fulfillments from our private points of view will enable our foreign visitor to see through the corporate shortcomings of a nation; and that he will somehow learn to love America as we do, in spite of herself.

Modern Provincial

A YOUNG FRIEND OF MINE WHO LATELY RETURNED from overseas with her Air Force husband remarked as we drove up a familiar road one day, "It's sad to come home from an experience that changes you so much and find that nothing else has changed."

She was expressing, I suppose, what every young person from the beginning of time has thought on his return home from "the great adventure." But I'm afraid I heard her calling us provincial, and I was defensive. "There comes a point in life when you take comfort in the fact that someplace things remain the same," I said.

We didn't pursue the subject. There was half-a-generation gap between us and we were at odds on point of view, but I think often of what I should have said. Each of us was justified in the emotion she expressed, but our facts were wrong. Life does change everywhere, even though on the surface things may look the same.

I presume that centuries ago the Crusaders who returned to their feudal estates in Europe from the Holy Lands looked about them and said, "Nothing has changed here." But, of course, they had changed so much themselves that eventually they changed the whole world.

I drive very frequently along the road where I grew up. On the vacant lots in our neighborhood many of the neighbors' children built homes and settled next to the folks. On a Sunday afternoon I notice in the yards and on the porches people who bear remarkable resemblances to adults who lived on that street when I was a child.

Acquaintances who see me in my mother's yard must have the same sensation, for I strongly resemble her. It would be easy to say, "Nothing changes along this street." There's a sort of blind security in thinking that. I no longer know all those people so I can't say for certain, but judging from my mother and me, I'm sure they've changed in proportion to the experiences they have had.

In his application to a university, one of my friend's sons wrote with a touch of wit, "I have grown up in a narrow-minded rural community. Now I would like the opportunity to submerge myself in an urban culture and form my own prejudices." He will go off to the city and synthesize a new wisdom, as well he should.

What, I wonder, will he do then? Will he come home, look around, and say, "Nothing ever changes here," then carry his wisdom off to a colony of seers who glory in wisdom for its own sake? There are colonies of such visionaries in all our large cities and clustered about every university, "fomenting change," they say. I have great respect for wisdom and vision. I'm sure it is easier to acquire and sustain in the company of the like-minded.

But change, I maintain, must take place back home among the "peasants." I know what it is to go home and try to retain a vision. It takes determination and constant renewal. It takes grit and a great deal of compassion (as anyone who has ever canvassed a community to promote a cause will affirm).

Perhaps what I should have said to the young woman disappointed in the community was, "We'll be expecting a fresh wind now that you're back."

8 CHANGE

To be on your way is to be home.

Joseph Pintauro, TO BELIEVE IN GOD

The changes in a way of life come as unobtrusively as the "For Sale" sign on the piece of scenic property down the road. The price is high and couples will dream over it awhile; but one day it will sell and an architectural marvel will arise, then another and another, and our little dead-end road will join the rest of the area as a gracious country-urban commuter address.

Surely farm life is on its way, as it has always been, to being something different. Those of us who are at home in the transition are the happy ones.

Evolution of a Breed

BROWSING IN A BOOKSTORE RECENTLY, I CAME across a photographic essay of *The American Country Woman* as photographed by Dorothea Lange. On the cover was the little old lady of the center part and the thousand wrinkles who seems to have symbolized the American farm woman for all of our three hundred years, the incarnation of Grant Wood's farm wife in "American Gothic." My immediate reaction was negative. I can love that little old lady, but I couldn't see her as representative of today's country woman.

But it's a subject I belong to, and I proceeded to turn the pages. Who ever buys these photographic essays, I wondered. This one was already dog-eared; a bookseller must expect that of a picture book. I read the index and the introduction. I met the first woman and the next and the next; each in her element. And gradually I moved back to a world of rickrack and worked buttonholes, of clapboard and parlors and mother-of-pearl pictures.

There was the house where I was born, the curtains at my windows; there was our church and its women turned out for cleaning day. There was my grandmother's garden, there the cupboards in our kitchen with the chips and dents we never saw.

And the women—my mother was there, my grandmother, and my great-grandmother. They were women of muscle and sinew, women round and sturdy, of sacrifice and duty, whose eyes had squinted too long into the sun. They weren't all happy women, but they were courageous and patient and kind. The text identified them; the photographs told their story.

"The land was ours before we were the land's," said Robert Frost in "The Gift Outright." These were women who at last belonged to the land and had found "salvation in surrender." My eyes were full of tears, and there was no indignation left in me—only pride in a great heritage.

These photographs had been made, I noted, between 1931 and 1953. Perhaps this portrayal couldn't have been done later than that. The American country woman today is difficult to distinguish from her counterpart in town.

The granddaughter of one of these women may wear a wig to conceal what a milking parlor does to a hairdo. She may lather with face cream against the ravages of wind and sun. The time her mother invested in worked buttonholes she may give to the League of Women Voters. On her walls she may hang impressionistic watercolors or sensitivity posters. Perhaps she plants bedding plants where her grandmother had such success with hollyhocks, and she may have discovered that she can't afford to preserve foods not grown on the place (or even those that are!).

Nonetheless, the country woman today knows as those women did that she belongs to the land along with her husband, that there is "salvation in surrender." She knows that a new combine takes precedence over a family room; that a farmer has only one hired hand he can absolutely depend on—her; that schedule is a fickle word; that weather is the greatest unknown next to income, and that everything depends on it. She understands that her children need a sense of the world, but that work never goes out of style. And she is aware that these will only bring security in combination with love and trust.

She may communicate more freely with her husband than either her mother or her grandmother did, but she has the same silent understanding of his moods, his needs, his despairs and joys. She will stand with him as he surveys a flattened grain field and weep for him as he weeps for the crop.

Life is better on the farm today, not because it makes fewer demands on an individual, but because the demands are of a different nature and the rewards are greater. We may never have etched in our faces the character that I found in Dorothea Lange's subjects, but I hope there will be a joy in us that the rigors of existence robbed from too many of them. (I bought the book, of course, dog-ears and all. There was too much of my life in there to leave it behind.)

The Hired Man

LIKE SO MANY OTHER FEATURES OF THE FARM today, the hired man has changed. The slightly dull, undependable, misplaced person type who slept in the attic, hid whiskey bottles in the barn, who had "nothing to look backward to with pride, nothing to look forward to with hope" * has passed along with the outhouse, the workhorse, and the hired girl.

It may well be that the liberated generation of farmers'

---

* Robert Frost, "Death of the Hired Man."

wives dealt the first blow to the system when, having demanded and won the right to have homes of their own free of maiden aunts and bachelor uncles, they raised a further howl about cooking and cleaning up after a hired man. But surely modern technology wielded the death blow.

There is still a place on the farm for the hired man, but today's man is an improved strain. (And stigmas being what they are, he won't be called the "hired man" but the "farm manager.") In addition to a strong back, he needs a sizable brain. To some degree he must be a botanist, chemist, agronomist, accountant, psychologist, time-study expert, mechanic, and methods engineer. To be truly effective he needs an objectivity that can rise above the entrenchments of tradition and sentiment.

In return he will command a standard of living at least as good as the farmer he serves. He will share the farmer's headaches and all his pleasures, save perhaps that of owning the land. He may, however, own a piece of property somewhere else that's turning him a fair profit.

Where does one find such a marvel? I'm of the opinion that you cannot find them; if you are extremely lucky, God will send one. If you have strong intelligent sons, perhaps you can rear one; but he may lack the objectivity you need. If he has it, you may lack the good sense to let him act upon it.

Our Ed came and offered himself to us, and we took him on his own terms. In his years with us he has earned a college degree, and now is an able schoolteacher who finds it convenient to farm summers and weekends. He is the soul of competence, but we find it disturbing to leave him here alone. In our absence he always has a host of insights that need implementing.

Another thing is true of our kind of hired man: He can get a job anywhere else at twice the money and half the hours. Of course, he wouldn't be happy, for once a man has surrendered himself to a love affair with a farm, he will ever yearn after it.

Sing a Song of Freeways

In order to appreciate the miracle of our superhighways, it helps to recall the awe with which we heralded the Pennsylvania Turnpike when it opened back in the 1940s. It was everybody's ambition to make the arduous trip to Pittsburgh and see this marvel that had come to pass, to take a ticket and speed forth in the super-safety of separated traffic lanes, released at last from the frustrations of cramped roads and stoplights and railroad tracks and pedestrians. Stopping at the plazas, truly elegant then in the eyes of middle-class Americans, we felt like wayfarers together on grand journeys into intrigue. Like Chaucer's pilgrims, every traveler had a story to tell over coffee at the lunch counter. No one was bound on a humdrum mission.

The glory of it all seeped into my blood, and I continue to be thrilled by the superhighway whether en route to the shopping center or outward bound to adventure. It gives me such a great sensation of America in its multiplicity and diversity.

I love driving through the tidy farmlands of western Ohio, knowing from my roots what life is like in those neat white houses in the corners of their sections, knowing what it is to be out on those tractors plying endlessly backward and forward, remembering what it was to be a yearning farm child looking at the highways and wondering if you would go any place ever.

I delight in the monolithic rows of silver high-tension towers stretching to the horizon, looking like Steinberg drawings of grand and priggish ladies. I have the feeling that they know that even headless they are the most powerful "women" in the country.

Sweeping up onto the elevated stretches of highway through cities floods me with compassion for the multitudes trapped there in "little boxes on the hillside" or in the somber dirty sameness of housing projects. But at twilight, city skylines are a special joy, when the lighted windows form patterns against the black buildings and the black buildings form patterns against the fading light.

I love silos, too, against the sunset; water towers like lofty shining pearls over small towns; the variety of church towers in aging suburbs that speak of foreign people far from home comforted only by their individual gods.

I like the sensual bombardment of traveling the freeways on a motorcycle, to feel the strength of the wind pushing against you, to smell the sweet scent of hayfields and vineyards and pine forests or the pungent odor of refineries and sewage plants and chemical works. Without the protective shell of steel and glass you are one with the elements. To the alienated youths hitchhiking near the exits you are Easy Rider, their brother, and they wave the peace sign. There is a youthful triumph in roaring past semis and buses and pampered people in luxury cars, knowing you share a secret they do not even realize exists.

To travel the freeways listening to the radio is to see and hear America simultaneously. Traveling through the poverty-ridden stretches of the South, one listens to gospel songs and mournful ballads of infidelity related in crippled grammar. You're left with the impression that their only hope is in the Lord, and you grasp the pain of the stunted lives bred in the tin-roofed shacks.

The screaming assault on the ear of the ghetto rock speaks of seething frustration with no hope. The recurring newscasts punctuate long trips in the same regular way as exits and restaurants and gasoline stations.

To travel the freeways is to be painfully, wonderfully aware of American industry and commerce, to be at once awed and revolted. There is strange beauty in the huge buildings with their superstructure of conduits and transoms and chimneys. But there is a brooding evil in the power they would seem to wield over the lives in the autos creeping in and out of their parking lots in snakey lines.

And there is always the persistent affront of the junkyards and the billboards and a variety of land pollutions.

I like going east into the rising sun, west with the sunset. I like the mysterious privacy of driving alone in the long night, the mind dreaming backward and forward into yesterday and tomorrow, not always being able to distinguish between them.

Yes, I love the freeways. I am a creature of this asphalt age. I love speed and being part of the power release. When I am driving on a superhighway with the cacophony of "radioland" blaring around me, I am one great lump of sentimentality about this good, bad, ugly, beautiful, terrible, wonderful America.

Fragrance of Honeysuckle

THE CURSE OF THE SUPERHIGHWAY IS THAT IT hurries one through and past life at a speed out of all proportion to life's value. Once upon a time, a trip through the Kentucky hills required that you wind up and down steep and narrow roads past mountain shacks and rural schools and general stores, through little towns and mission stations and national parks, stopping occasionally for food or rest or directions, talking with curious natives whose days your visits sparked with something out of the routine.

You came away with the feeling that you had been someplace else, that you had an understanding, however limited, of a kindly, straightforward people. Their virtue was that no matter to what degree life had passed them by, they at least had not bypassed life.

It was the fragrance of honeysuckle pervading all the Kentucky valleys one June morning that seductively lured my sister-in-law and me away from the interminable concrete, away from the world of air-conditioned cars and gas stations and Holiday Inns. Setting off on foot, we found a slag road that led along a dry creek bed in a valley, past weathered cottages where children and chickens and hound dogs mingled in dirt yards, where washing done on the front porch grayed discouragingly with the dust of the road beneath tulip and sycamore trees.

Down a narrow path to the road came two early-morning children scrubbed and neat. Shuffling briskly along behind was their arthritic grandmother leaning on a gnarled walking stick. Her hair was caught up in a knot and her skirt was maxi, revealing only a few inches of ankle above worn tennis shoes. I had the feeling that they had been sent by a special department of local color.

We passed a small field of potatoes, which at a distance looked prosperous out of all proportion to its setting. On the porch of a house set uphill from the potatoes, two old women rocked in ladder-back chairs. I waved and called, "Nice potatoes. Yours?"

"The old man's," one replied, indicating a field beyond the house where a figure bent over a hoe. "Where 'ya goin'?"

"Up the mountain to pick flowers," I said, indicating the clump of honeysuckle I was already clutching. But it seemed unneighborly to pass the man in the field without chatting. We made our way through the honeysuckle hedge to where he worked over his tomato plants.

"Nice garden," I said, admiring the peppers and cabbage, the tomatoes and the corn stalks at the base of which were planted pole beans.

One wondered if there was nourishment enough to sustain both, wondered which day a "gulley washer" might flood his total effort down the steep incline to the road and resign even this plodder to "what's the use?"

His potato patch was clearly his pride, already a foot high at the end of May.

"You must have planted those back in March," I said, commenting that my husband was a potato farmer.

"Nope, middle of April," he said as he spat proudly. "Gotta' keep a lot of that there sodium 'longside the plants and wet it in."

"We got our early ones in around the first of April," I said, "but they're not nearly that high."

"Well," he said, dismissing the subject quite simply, "maybe you'uns ain't 'farmers.' "

A closer look at the patch revealed that his foliage was thick with flea beetle damage, and I sighed inwardly for the disappointment ahead.

He walked us down to the road past his poor but neatly fenced homestead. "There," he said, indicating a weed growing beside the fence. "Do you know what that is?" We didn't, and he went on to explain.

"Them college kids dry that and smoke it and it makes them drunk."

"Marijuana?" I asked, incredulous.

"I don't know what you call it, but it makes you drunk when you smoke it," he repeated.

We said goodbye, and as we walked on along the foot of the mountain, I told Alice that I thought his future could be bigger in "grass" than in potatoes.

The road led through an abandoned sawmill where stacks of lumber turned gray and weathered in the weeds. All along the route, both near the cabins and remote from them, were derelict automobiles, seeming rejects from a society not yet ready to cope with them. Those of longer standing had become enveloped and obscured by Virginia creeper and honeysuckle. Daisies grew up through holes rusted in fenders, and saplings pushed their way through bared chassis.

Our road became finally no more than an old logging lane winding tortuously and delightfully up the mountain through the forest. Alice, who is an amateur botanist, reacted jubilantly to each distinct specimen of plant life, familiar or not. The daisies were enough to satisfy me, but I added bits of color to my nosegay with hawkweed and heliotrope, wild rose and myrtle.

I would have abandoned the climb as the freshness of early morning gave way to the heat of 9:30 and gone back to sit on one of those cabin porches, but Alice had that spirit of "Excelsior!" We trudged upward, running finally into pine and laurel; stopping to admire mosses and lichen; to listen for bird calls; to collect along with our flowers a gnarled root, a piece of bark, and the sight of a new bird.

Alice was optimistic that each turn would produce the summit. When finally we had achieved a panorama of valley and our lane forked, forcing us to make a choice, I stepped off to the side to peer through the trees toward the valley. There below the roadbed, nestled in the roots of an oak, was a sprawl of azalea blooming in a flame of brilliant orange.

As the climax of a quest, botanical or aesthetic, this was sufficient. It called to mind Gray's metaphor from "Elegy in a Country Church Yard" of the flower (and the girl) "born to blush unseen and waste its sweetness on the desert air." Would anyone else ever find it there, that flamboyant flower . . . or any of the other colorful "flowers" we had discovered that morning—the old woman with her children, the old man with his hope, the gossips on the porch? Or will they all perish forgotten in the weeds that grow up alongside the river of concrete?

The Predators

MY FAVORITE CHRISTMAS PRESENT THIS YEAR was a simple can opener. Yet I have pangs about the little "gem" it replaces. I remain very attached to old things, and especially this can opener, for it was the only mechanical object in the house that I understood and operated better than the men. (Paul thought it was a "beast," though that's not the way he phrased it!)

My sons thought I should have an electric can opener; however, the things terrify me, and I threatened a kitchen boycott. Perhaps it's that old dental-drill mentality, but I really think it's something more primitive. I have a fear that they might reach out and grab me.

Electric can openers aren't the only gadgets that make me wary. I have always felt that revolving doors were out to get me (from the rear!). And I certainly don't trust those doors at the supermarket that promise to open when you step on the magic rubber mat!

Aerosol cans! There's an abomination! I haven't felt comfortable with them since first I garnished the ceiling with whipped cream. I resent the fact that the pressure is usually exhausted before the contents. And can anyone ever be really sure that he won't get spray paint in his face or shaving cream on the mirror? Who can trust a kid when there's a single aerosol can of anything in the house? Besides, I resent paying so highly for what I enjoy not at all.

Only worse than aerosol cans are vending machines. Parting with money has always been painful, but dropping it into a slot for who knows what (if anything) is sheer agony. I die a little each time I relinquish a coin. My children have learned to profit from other people's mistakes, and can usually get more from a vending machine than they pay for. I don't know how, but I'm sure it's legal. What is illegal is that I only get half of what I pay for.

I can get up in an auditorium and address a crowd of people with relative calm, but just connect me to a telephone recording machine and I freeze. Even a Xerox machine gives me the creeps, and electric typewriters—ditto!

We have a toaster that "snatches" the toast and pulls it down, then releases it in a slow, sly fashion. I'm terribly suspicious of that thing.

I have never had occasion to work a computer, and it's just as well. And just thinking about a garage door that might go up or down when I wasn't around to will it so is enough to awaken me in the night screaming!

There is probably one of those long "phobia" names for what is bugging me (techniphobia?); but, as I said, I think it's more primitive than phobia. My understanding of evolution is that living things either succeed in developing a defense against their predators or they become extinct. There's little question in my mind about what the predators are. I shall continue to be wary, and I shall grow more so . . . and I shall continue to rejoice in my very simple can opener.

Happiness Is a Human Voice

STORIES ABOUT PEOPLE WHO HAVE TRIED TO establish contact with computers are legion. The best one I ever heard was of the man who cut a filigree of those little window-shaped holes in his "don't fold, don't spindle, don't mutilate" card spelling out the word "cancel." He made contact!

It was completely by accident that I learned how to make a computer light up and say "Tilt!" All you have to do is record your Social Security number incorrectly, provided, of course, that you have previously recorded it properly. This is not an unlikely circumstance, considering that you pass out your Social Security number almost as often as your name.

Whether or not one has ever tried to "get through" to a computer, he feels somehow victimized by a hostile machine each time the mail delivers one of those rectangular cards with a corner missing. What is it about them that engenders hostilities? On the reverse side there are usually printed a few hundred words of legal jargon so small as to remind one that he's middle-aged. The subject of the card itself, usually representing payroll deductions or monies otherwise extracted—or to be extracted—is surely sufficient to inspire unkind thoughts. But the subliminal antipathy I sense more than any other is that the IBM card is *not* the warm, friendly greeting I never cease to hope I will encounter in my mailbox. Even a check cranked out by IBM lacks the warmth of the hand-produced job.

So—imagine my delight on a recent evening in picking up

my phone, acknowledging my name in all its pompous length ("Yes, this is Patricia Ryall Penton Leimbach") to find myself connected to the man behind the computer down in the state capital.

"You're who? C'mon man, you're putting me on! It's way past five o'clock, and you don't sound anything like a computer."

His name was Bernard, and he sounded like a guy who might love his kids and go bowling and even be tempted to mutilate an IBM card! "Say, Bernard," I said after we had straightened out the discrepancies in the Social Security numbers, "as long as I'm talking to you, would you mind answering a few of the questions that have been bugging me through the years about this account?"

"Shoot," he said, and I fired my questions. Out of his vast memory bank he drew the facts and figures that had eluded me so long. When I didn't "read" him, he repeated and explained patiently.

"You've been so helpful, Bernard. I can't tell you what this means to me."

"That's perfectly all right," he said. "You've got my name there. Don't hesitate to call if you have any more questions. And if you know anyone else who has problems about this, refer them to me."

And then he was gone. I sat there dazedly, wondering if I'd dreamed it all. But no, my little book was open to the page with the Social Security number. The next morning I called the near end of that bureaucracy, and sure enough the error was real.

Ah, Bernard, you devil you, sneaking back to the office at night to finish off the day's rejects, making your long-distance calls on the after-six rates. You restore my faith in man (versus machine). You have put a smile in my life, a name on a computer card, a face on a vast bureaucracy—you have seduced me with your voice!

Rape

At the top of the hill stood the house, square and green-shingled, looking down the street with big white-curtained eyes. A low hedge surrounding the steep yard bespoke one man's effort to put a circlet of order and love around the home he established there long ago.

It was the last day of the year, a chilly day with a threat of snow, a day suited to the task of clearing out a house, whisking away a lifetime of accumulated treasure. Widow Murphey came back with her son and her nieces to preside over the dissembling, not because she could really help or because she wanted to come, but because she couldn't bear not to come.

Time had long ago ceased to move in this Victorian house, probably ten years ago when "Pater" Murphey died. The walls were dingy and the furnishings out of date. What was the use of changing things? This is the way she had shared the house with the people she loved, and the warm memories were closer when the house stayed the same. The house and the furnishings had never been central to Grandma Murphey's concern; only the people had mattered, and they were gone. Yet in terms of life invested, each item here had value.

The dissemblers rushed through; time was short. It wouldn't do to be caught in these Pittsburgh hills in a snowstorm with a truckload of furniture. No time for the loving pondering an old lady would appreciate. The big decisions had been made beforehand, but there were hundreds of small ones to be made today hurriedly, without sentiment.

Grandma Murphey wandered about steadying herself on the furniture, the woodwork—the handholds she knew so well. She was confused by the activity of so many well-meaning intruders rifling through cupboards and drawers and closets making "no-value" judgments on so many things so precious—recipe files, an old sewing box, a drawer full of aprons, the Fiesta kitchenware so lovingly preserved.

The minister came and was introduced. The old lady had hoped he would take her piano, but no, he had no need of it. He appeased her by making a selection of books, books of theologies as out of date as "Pater" Murphey's chemistry texts that shared the shelves.

The men hauled her bedroom set downstairs and out to the truck. They rolled up the rug, carried out small tables and her wicker fern stands These things she would want in her new home with her son.

The women continued their plunder, wrapping the crystal and china, dividing the items of worth. The curtains they would leave for the new tenants. Perhaps some of the furniture could be sold; the rest . . . well, maybe the Goodwill. . . . With Mrs. Murphey close at hand, her daughter-in-law came upon a flat book labeled "Memories," easily recognized as a funeral parlor registry. She quickly thrust it back in the drawer where she had found it, lest any of those "memories" escape on this day when so many memories were already at large.

Neighbors hurried in to say goodbye and embrace the frail little woman, trying their best to bring cheer to a cheerless occasion. "Hell," said the rough old codger from next door, "I ain't gonna' say goodbye—just so long!"

The boxes had gone out, and many items of afterthought piled on top. The truck and the several cars were full. The dissemblers hurried about double-checking the attic, the cellar, the garage. The old woman stood alone in the living room. "Who will take my piano?" she asked of no one in particular. "I've had it since I was fourteen." What an event it must have been when this wondrous extravagance came to the tiny Ohio town where Mrs. Murphey grew up. Small wonder she treasured it.

She moved in her uncertain steps to the piano bench, put her fingers to the keys, and played, stiffly and slowly, but with a memory that was fresh, a few measures of "In the Gloaming." Then she switched to some bars of "Lilli Marlene." The heart of that home beat faster in those few minutes of an old woman's music, and then failed.

They found her coat and scarf, helped her into them, then led her out of the house, down the steps, through the hedge. As the car carrying her disappeared down the street, the house stared after with big white-curtained, now soulless eyes.

Case Study

IF YOU WANT TO STIR UP A HORNETS' NEST, JUST bring up the subject of Women's Liberation at the neighborhood coffee klatch. When the movement started—years ago—I already considered myself among the liberated. Yet I must admit to a short-sighted sense of amusement at the hullabaloo militant feminists were raising. Mine was a quiet revolution—and only as a result of reading and reflection do I now realize it took place.

The idea of female inferiority is ingrained in our culture, sometimes in subtle and seemingly insignificant ways. Inequalities range from property rights to business, professional, and political opportunities; from sex symbol debasement to attitudes of superiority that have caused many women to accept the status of second-class humans.

Perhaps my case is extreme. I was born into a family of men. I married into a family of men, and I am rearing a family of men. Logic says that my rage at the basic inequalities between the sexes should be directed at men. But for me that is almost impossible.

I am surrounded by men who love and respect me and are as innocent of malevolent intent as my mother, who tried her darndest to make me a paragon of womanhood.

And, in the eyes of the world at least, I am a happy, satisfied, reasonably independent woman.

I have labored toward liberation through twenty years of marriage. I was born into a family of five boys, one girl. For Father, paternal dominance was an assumption enforced by screaming silence. His bigoted insistence on it was almost the only source of unhappiness in our seemingly well-ordered home.

My sister and I grew up never questioning the "fact" that male word was law.

When I married, I moved into a situation of male dominance more destructive of my identity than the one I had left. My mother-in-law was a strong, capable woman who could have built and ruled an empire had she not submitted to the prevailing opinion that woman's place was subservient to man. I feel certain she died an unfulfilled human being in spite of a string of laudable accomplishments.

From her example, I learned not to buck the tide—I submitted completely to being a farm wife. I went to the fields; did what was asked of me always. I drove a truck and did chores when called upon, and learned to enjoy the work. Though I often seethed in disagreement, I never tried to intercede in the decision-making between my husband and father-in-law—they weren't conditioned to respect the opinions of women in their business.

Then, gradually, I began to need help in the house to compensate for time spent in the fields, time off to attend conferences or retreats, to take part in activities where bit by bit I unearthed my identity as a person beyond wife, mother, and farmhand.

My husband is a gentle man, well-schooled in the household arts by that mother of three sons.

We came to share responsibilities for the house, the farm, and the children. I grew to understand that if I thought of myself as chattel it was a preconception I alone harbored; he had outgrown it.

Considering the happy balance of responsibility that he and I have achieved through twenty years, I see that "women's liberation" is a misnomer. What really is involved is *human* liberation. When women cease to be shrinking violets, they unburden men of a great deal of responsibility, both foolish and real.

Perhaps it is true that farm women as a class feel more liberated than others. While many farmers do realize the true economic value of working with their wives as partners, I know a lot of farm women who feel that the wrath of God would descend upon them if they did not have lunch on the table at the stroke of noon.

A farmer's infringements upon his wife's time and schedule are seldom questioned, but don't ask him to rattle the pots and pans. A wife gets little consideration for a work-day extended by farm responsibilities until bedtime, while her husband usually relaxes after chores. On an economic level, his ten hours may be of more value; on a human level, she is left just as weary.

For all their quiet submission, these are not unhappy women. Many of them, like my neighbors of the coffee klatch, are enraged by talk of Women's Lib.

Men who think Women's Lib is a big joke have no conception of how much more wholesome and fulfilling a marriage can be when a woman shoulders her share of the big problems and decisions, and he takes on a share of the small ones.

It is not enough for husband and wife to "do their own thing." They have to learn to do each other's thing. And a liberated woman is going to end up doing a lot more of "his thing" than she had expected! But experience has also taught me that when husband and wife operate on equal footing, life can be richer, fuller, more satisfying.

Surely our children do not see their father as less of a man because he can iron his own shirts and sometimes does; nor do they think me less of a woman because I go off on a potato delivery with a two-ton truck which I am capable of unloading. I think they see us as a team working for mutual satisfaction.

I do not advocate a reversal of roles or even a realignment in situations where freedom and harmony obviously exist, but I feel every marriage should be open to constant reappraisal by the parties involved. Whether we approve the trend or not, it's a fact that nearly fifty percent of American women work outside the home. When a couple agrees that she should go out to work, it follows that they should share the responsibilities for physical care of house and family. Too often, children are losers in the equality business.

If I am in fact a "liberated woman," what then is all the beefing about? Perhaps it is directed hit-and-miss at fate—at culture, religion, and society in general—for perpetrating the myth of inequality.

I suspect much of the anguish is directed inward. The scars are within *me*—so deep that I can never *fully* accept the truth that woman is not less than mankind.

I weep that I have no daughters to whom I can pass the truth. But, oh, my sons shall know!

The Green Machine

TEDDY'S GRADE-SCHOOL ART HAS BEEN ONE LONG parade of farm machinery—tractors and cultipackers, tractors and potato planters, tractors and sprayers, tractors and corn pickers, everything that moves on wheels, in its season. And one fall there emerged from the drawing board a great green monster of irregular shape spewing potatoes from a mouth at the end of a long neck, towed by the familiar red tractor, drawn a bit larger this time.

Square purple people were riding this machine, and if like some avant-garde creation from the Museum of Modern Art the picture had been wired for sound, it surely would have passed for the old "purple people eater." Philosophically considered, it is a people "eater." This money-green machine has replaced ten or fifteen potato pickers who came on their off shift, their off days, between jobs to carry out the potato harvest.

I miss them, those friendly people; I miss their banter and good cheer. I miss the sense of oneness we achieved, black and white, farmers and laborers, men and women working together. I miss the friendships that grew out of those numerous encounters.

True, people ride this machine, but they too are swallowed in its machinations. Gears mesh, the power take-off engages, the squeaking and squealing commence like fingernails scraping a blackboard, people shift into motion moving stones and clods and potatoes—chasing, chasing, chasing along the moving web. Conversation is stifled and an air of the frantic prevails. The machine stops, and as though they were wired into it, the people stop too, sinking exhausted to the railings behind them.

But, wonderful green machine, it does the job—quickly, thoroughly, and (mortgage payments excepted) cheaply. The quality of this herbivorous monster is that, unlike the hit-and-miss crews of humans it displaces, it is dependable. It sleeps in the barn where nineteenth-century posts were moved to accommodate its twentieth-century contours; and when the farmer wills, it bites into the soil spitting potatoes to the side, excreting weeds and vines to rear, burping stones and clods.

No more looking down the road of a morning and wondering if anyone will show, no more looking to heaven of November dusks saying sadly, "Looks like a freeze tonight and twenty acres still in the ground." No more long afternoons of working on our knees, Paul and I, finishing the picking no one else had time for.

It was inevitable, this device that makes the potato farmer more independent of the vagaries of human nature. A twentieth-century America could not exist on the production of a nineteenth-century threshing crew. And like them, the potato-picking crews will vanish. But never since the combine, the baler, and the hay chopper, has the same strong feeling of mutual concern existed among neighboring farm families. Every device that fosters independence fosters loneliness and turns man inward.

9 RENEWAL

It is not really rest or even leisure we chase. We strain to renew our capacity for wonder, to shock ourselves into astonishment once again. *Shana Alexander,* "THE ROMAN ASTONISHMENT," LIFE MAGAZINE

Renewal is as quiet an incident as a walk to the spring in search of skunk cabbage in early March, or as WOW! *as a ski trip abroad. It may be the joyous advent of a house guest, or a hectic day at an amusement park. The essential ingredient in every case is a perspective that provides insight on the everyday and a greater awareness of life.*

Orrin is young enough to see truth in a straight line. He says simply, "Home is the best place."

"How did you know?" I ask.

"By going away."

Renewal

ONE OF THE NOURISHING RITUALS OF MY LIFE IS an annual retreat for church women where I meet stimulating people and live richly and deeply for three days. There I have found a different "me" who broadens the viewpoint of the farm wife. Not the least of the joys afforded by this privilege made possible by an understanding husband is the great pleasure of coming home.

In an enumeration of things that enchant, J. B. Priestly wrote, "The beginnings of journeys are delightful. . . . But better still is coming home after a long absence, the moment you open the door."

The delight of "coming home" is not now what it was when I copied those lines in my diary. As a college freshman, "coming home" was the strangely inverse growth from small fish to big fish, a liberation from a phony front, a return to security.

"Coming home" as a wife and mother is just as gratifying but in a different way. The central fact in each case is that home is the place of all places—here life is most meaningful, and this is best appreciated by removing oneself from it occasionally.

Going away with the family is for the most part to take "home" along and to come back tired and peevish. The woman who goes away alone and returns recognizes that here among the "things" of her married life, in this home she has created, she does not seek security so much as she supplies it. And in her absence this fact has been appreciated by the other family members.

There is an aura of idealized detachment in the return view. This is a place that exists only following your absence. Now you view the uncleared walks, the cluttered porch, and you couldn't care less. The living room with three days' papers, the dining room with its homework confusion, the kitchen with crumbs and garbage, the bathroom with scattered laundry—all are symbolic of a family's need of you, but not so much so as the look in their eyes.

As the absent three days unfold in the children's subdued and urgent conversation, you realize how they leaned on each other for sustenance. At bedtime they linger, not their usual dalliance but a genuine pleasure in having you home. All patience and forgiveness, you see each one for what he is and for the extent of his needs.

Then, in the after-bedtime peace, you sit with your husband at the dining room table and fill in the small details, recognizing that a life of love is built on a rock of compressed minutiae. Tomorrow you can take up a vocation that often seems stifling and futile with renewed enthusiasm and new vision.

Winter "Idle"

MY FAVORITE WINTER TREAT IS BEING SNOWbound. There was a time when here at End O' Way we could depend on being snowed in at least two or three days a winter. On one frivolous occasion a Leimbach Saturday night party went on until the snowplow arrived Sunday afternoon. Baby-sitting grandparents burned up the telephone wires with disgusted "I told you so's."

But more efficient snow removal equipment has made a repetition of that folly more and more improbable. Now and again, however, comes the pleasing word over the airwaves that tomorrow as previously conceived has been cancelled, and a new tomorrow must be shaped.

Being snowbound on the farm in this age of electronic communication is certainly not what it was when John Greenleaf Whittier wrote his idyll on the subject in 1865, but it does involve some of the same satisfactions—unexpected family time together, pride in challenging survival, a generous measure of spectacular beauty.

Snowbound is cozy. Everybody is home (a rare circumstance) and nothing is planned. It's a good time to sleep in, bake bread, start a Monopoly game, make soup, play records, make ice cream from snow, read "Snowbound" aloud, work on the income tax, pop corn, sit by the fireplace, read a good book, take a nap, write Christmas thank-you's, organize the slide collection, make scrapbooks, have pillow fights, observe afternoon tea . . . well, the list of home-grown pleasantries is endless.

When the snow ceases and the wind subsides there must be outdoor sport to mark this as something more than a routine snowfall. If you have a turn-of-the-century barn filled with rustic antiques, and the farmer can be coerced into dragging out the ancient bobsled, you can create your own episode from "Snow-bound," as we did this winter.

"If Grampa were alive, I know what he'd be muttering,"

called Dane to me from where he sat on the tractor. "Damned foolishness!"

"Your father's very words!" I called back, as I wrapped the little boys in a blanket and we took off up the snow-covered lane.

Being snowbound is something more than the sum of the joys and rigors it involves. It is a blessed retreat from frantic lives of perpetual motion. In the middle of one snowbound night I stood at an upstairs window and looked out on the white peace that isolated us. I thought with warmth of the solidarity born of isolation, of how much these two unscheduled days had meant to our family.

As I stood there wishing we might be cast back into Whittier's day and be snowbound a week or two, there came a rumbling, a flashing of yellow lights, a flying cloud of snow, and the snowplow thundered past.

New Hampshire Holiday

THE CATS CLUSTER ABOUT THE DOORSTEP AS WE let ourselves out in the subzero air. Teddy crosses the drifts to break a five-foot icicle from the low eaves. Snow crunches and everyone steams as we pack our gear and set off for the ski hills. The roads are black channels carved in white valleys. To help justify the children's week away from school, we do a geography unit on New England as we cover the miles. "These are the White Mountains way to the north. Notice the rock outcropping. That's typical New Hampshire. See how the barns are built onto the houses, on account of the snow. And notice how steep the roofs are. What kind of farming do you suppose they do here? There's an old mill with a working waterwheel. What do you suppose that does? And why do they have so many streams up here? Where do you suppose they're going with that load of logs?" The distance is covered in discovery and wonderment.

A ski day is a sensational day from start to finish. There

is the unparalleled sensation of standing atop a mountain and marveling that (1) you arrived there alive (2) such beauty exists; then the fright of a first run down on too crusty snow.

Camaraderie hangs in the air like your breath. The totality of the common struggle unites all skiers— "all of us" against the rigors of the sport. The children are a marvel enjoyed by everyone—on the trails, on the lifts and in the lodge.

"That's the time to teach them," they call after a parent, and a child's welfare is everyone's concern. Over an intermission coffee I watch the tiny form of Teddy appear untimely at the brow of an expert slope. He has taken a wrong turn and ended there where he has no business to be—flesh of my flesh, hesitating, groping there beyond my influence. I am suddenly struck by the insanity of getting him into such a predicament, yet I know as surely as he'll map out his life, he'll map a descent, and he does. Cautiously he picks his turn, sliding on his bottom over icy patches too steep to ski, catching himself on loose snow, regaining confidence and form, and bombing down again to the lift. A truly strange sensation.

The sun shines and disappears; snow starts quietly, rests lightly on the gay woolens that pass. Paul and I stop with Orrin beside a grove of white birch to enjoy the beauty of this secluded moment—the quiet snow, the slim white trees, the happy child and happy us. We know what it has cost in money and time and distance and effort to be here, and it seems little enough to pay.

Karneval In Luxembourg

IT IS ALWAYS DEPRESSING TO ARRIVE IN A strange place at night with no one to meet you, and so after fifteen weary flying hours (including unscheduled delays) the Luxembourg air terminal was a cheerless place. As we claimed our VW from Hertz-Luxembourg and started down the narrow snowbanked avenue toward a hotel in town, we wondered silently why we had elected to make this crazy trip.

But within an hour we had found a room, a meal, and a couple of promising friends and were off to a masked ball au Pole du Nord (North Pole) answering our own question. *This* is why we came—it's Karneval!

The "friends" we met in the hotel dining room were an American professor and his wife on sabbatical from the University of Arizona. Their big problem was warmth. It's difficult to buy in Tucson the clothing one needs to be fortified for northern Europe in winter. The Pole du Nord proved to be a "hot" place, and in half an hour the tempo of the musique, the gaieté of the people, and the effect of the vin blanc had banished the cold from their minds.

Luxembourg had to be the best place in Europe for these four Americans, one speaking French, one German, one Spanish, and one presuming to speak a maxi language including all three.

The professor had been in France with the liberation forces and claimed to know "bedroom" French. A few minutes with him persuaded me that this was sheer bravado. (You have to be suspicious of somebody whose wife gives him a Key Club membership for his birthday.) He rationalized that living in the Southwest with the Spanish influence had corrupted his French. *He* certainly corrupted both languages, and every time he opened his mouth I winced. It took us only a few minutes to join tables with a group of Spaniards and Portuguese working in this tiny country. Our common bond was being "auslanders."

The dancing was most exuberant. It must be acknowledged, I suppose, that Luxembourg is a provincial sort of place, but it was refreshing to be among young people who dance together rather than passing the evening in isolated gyrations. Most of the people wore a semblance of costumes and masks, an element of anonymity that serves to break the barriers between people faster. I wanted to wear my ski mask, my flannel nightgown, and carry my hot-water bottle, but Paul wasn't in the mood for foolishness so we went as ourselves. As the evening wore on and the wine flowed, the conviviality passed even the peak of a good Polish wedding.

Delightful to be again with people who sing! We didn't always know the words but we could join in lustily on the Ho-lo-re-a-ro's, and even a high hum is a satisfying outlet. American music is always popular in Europe, and this multilingual group sang some of it in English, some in German, some in French, and some in all three at once. Humming along to a familiar tune, trying to catch the words, we were surprised to recognize "Comin' 'round the mountain" (Und wir schlachten den roten Hahn when she comes).

We hadn't slept for thirty-six hours, so with much cheering and handshaking we bid a gay bonne nacht (language was really polyglot by then) at 2 AM and went "home" to the eiderdowns at the Grand Hotel Cravat.

As for der Herr Professeur, a masked jeune fille had asked him to dance and he was three feet off the floor. His Frau won the door prize (a magnificent cactus plant—coals to Newcastle), and when we left he was ordering "catra beers." They had to catch an early train and we didn't see them again, but I trembled to think what shape they'd be in by the time they left Europe in May. Obviously warmth was ceasing to be the problem.

Ski Life in Austria

"HIGH SEASON" IN AUSTRIA MEANS THAT THE German population has migrated to the Alps for winter holiday, and we feel part of the high. This postcard Tyrolean village is swarming with robust individuals in stretch pants and fur coats who get their kicks by climbing the mountains on foot (often in those fur coats), renting beach chairs, and sitting in the sun. The less vigorous take the ski lifts up, toting a grandchild or a blanket or a bag with lunch. Around three o'clock they mosey villageward to take tea at one of the hotels or guest houses. It's a great family affair, winter holiday, and has all the earmarks of summer at the shore, including the tans, many of which are too deep to flatter.

There are plenty of skiers of all ages, but sport is not limited to skiing and hiking. Many people pull one of the short, high wooden sleds to the top and slide down in numerous installments, braking with their feet. Others take their lives in their hands with a contraption called a ski bob, a seat mounted on a ski (again with feet for brakes). Suffice it to say that with pedestrians and sleds and ski bobs, skiing offers more hazards than usual!

For several days we were in a ski class, for ski lessons at $9.60 a week are difficult to resist. Sigmund, the instructor, did more for my skiing in a morning than I've accomplished alone in ten years. Every turn I made down the hill he hailed with, "Bravo! Prima!" (Paul insists it was only because he didn't know any critical words in English, but I ignore him.)

I continued to feel "prima" until one of the less prima pupils ran into me as I stood on a ridge about two miles up the mountain, knocking me downhill about fifty meters. (Another of those extra-skiing hazards.) I've learned how to say "black-and-blue bruises" in German.

After two days at a stuffy hotel in the village I went looking for something more friendly (Paul hates that "looking" business). With fool's luck I stumbled upon Haus Talblick, an extremely gemütlich guest house "up the road a piece." (I suppose it was the little children playing in the yard that attracted me.)

For $2.00 a day Frau Taller provided bed and breakfast and tender loving care to twelve guests. Her housekeeping, like her smile, is "prima"; the rooms are warm, the water's hot, the beds are down (as opposed to feathers). The toilet, of course, is across the hall, and the window is always open. (Europeans, we conclude, have a permanent "outhouse" complex.) Bath is by appointment.

The dining room is the social center of the house, where everyone gathers before-ski, aprés-ski, and into the night. We are apparently the only Amerikanishers in this village of 10,000 tourists, and we're not feeling the slightest bit "ugly."

The lone Dutchman in our group interprets whenever we don't understand the conversation. He is the life of the party, holding court for as long as the guests or the wine or the stories last.

Tonight the house guests toasted Paul on his birthday in three languages, finishing with "Hoch soll er lebe" (high shall he live). It was a day on which we walked up beyond the ski tows a mile or more to the peak of the local mountains to look over the gingerbread tops of half of Austria. The sun was warm, the sky bright blue, the snow crystalline, the air invigorating, the view—well, words don't describe it.

Remembering the exhilaration of standing on that peak at the foot of the inevitable cross that graces all these mountains, I can only concur that the birthday wish is a good one—we live very "high" indeed!

Barbie From Bobbie Lane

MANY OF THE FOOTLOOSE AMERICAN KIDS bumming around Europe and points south or east have one secure item tucked away in their passport folders: a return ticket on Icelandic Airlines. It was interesting to talk with some of them on a flight home from Europe recently and find out what finally motivated them to use those tickets.

Some had run out of money, of course; for the ski bums the season was practically over; one went because she was eager to be again with her family; but Barb Springer seemed to be going home because she had grown up.

She had left the house on Bobbie Lane in San Miguel,

California, last June—with her knapsack, her Girl Scout shoes, and her illusions about the grand and untrammeled world "outside." She carried along a burden of disillusionment too—with adults and their false values, with money and the way it twists people, with her contemporaries for their juvenile attitudes and antics.

We sat across the aisle from one another, and trying very hard not to seem judgmental, I questioned her: "Well, you've been on your own for nine months now. What have you learned?"

"It would take me two years to tell you."

"See if you can cram it into two hours," I said.

"For one thing, I learned that it isn't all as romantic and exciting as it seemed—junketing about on the loose. I've learned how really low and rotten people can be at times, the jams you can get yourself into!" (Barbie looked like a good strong girl and I didn't question her capacity for getting herself out of a jam.)

Every point she made involved us in long digressions, so it was a long time before I drew her back to the initial question.

"Another thing I learned was to appreciate my parents. I used to find fault with everything they did; like, before I left home I was always critical of my mother for being so poorly groomed. I've decided now that it's a pretty petty and unimportant thing." (Her blue jeans, her field jacket, and her unkempt hair made the point rather clearly. Of course, it's a long time between showers when you're hitchhiking the hostel circuit, and a backpack doesn't do much for clothes. I could easily forgive her.) "I still think my dad is obsessed with his job to the exclusion of everything else. He hasn't any time for his kids or the pleasures of life."

We got off onto parents, and I attempted to persuade her to consider what made her parents the way they are. She thought it would be nice if her parents were more like Paul and me. I pointed out that if that had been the case she'd have been home "tending the store" like our seventeen-year-old, and her parents would have been out vagabonding!

The third lesson she claimed to have learned was the "value of money." She had evidently been without it enough to have discovered hunger. She stretched her dollars as far as she could, then took a job at something. There was certainly no blank check from home. She told of working two months to earn enough extra to send her little sister a Spanish guitar, and being disappointed at her sister's lukewarm response. "I feel sorry for my sister. She gets everything too easily, and doesn't appreciate any of it." I tried to assure her that the time would come when her sister would recognize how much love the guitar represented.

"And I sure found out how much I don't know!"—the beginning of wisdom, someone has said, though I really think Barb was born wise. She was friendly and candid, as honest a Girl Scout as anyone ever bought cookies from. I'm sure that wherever she had been she represented us well in her high-school Spanish. She was going home and back to school.

Our conversation terminated about the time we got over the continent. Barb looked out the window at the wastelands of the north, and with teenage enthusiasm shouted, "Beautiful America, and hamburgers, and p.b.j."

"P.b.j.?" I asked.

"Peanut butter and jelly!"

A girl like that can't be all bad!

Once she arrived home and "settled in," Barb was probably wiser—and sadder. She must have discovered that things were much the same—too much money still does ugly things to people; her contemporaries would only have grown nine months older; whereas she once thought her mother a mess, now her mother probably thought her a mess; and it's terribly difficult to discipline yourself to family living and study when you've listed with the winds for nine months.

But I wish you well, Barbie from Bobbie Lane. Keep

remembering the old Scout motto, "Be prepared!" I hope you get back to Spain and Portugal, to Morocco and Marrakech. Perhaps then you'll learn a couple of other truths: people are usually as fine as you expect them to be, and no finer; the world of vagabonding really can be romantic and exciting when you bring a few years of hard work and anticipation and preparation to the privilege.

Rock in the River

For those young mothers with a tendency to despair, I would point out that there comes now and then a day when the projects are caught up. I have, for example, mended the rips in next winter's jackets, and yesterday I finished last year's overalls. I don't care to dig much deeper than that.

This sort of miracle happens to farmers' wives when it has been raining for three or four weeks on end and working the land is out of the question.

So I sit on a rock in the middle of the Vermilion River relaxing insofar as it's possible for a farmer's wife to relax on a Thursday afternoon in July. The sense of guilt is as much a part of me as the restricted suntan I'm accustomed to acquiring bent over a hoe. The word "leisure" has never had much application in the lives of farm people. It was a word more often associated with the idle rich than the frugal peasantry. Surely we are among the last inheritors of the work ethic.

But this rock has lain here temptingly all my twenty-one years at End O' Way, and never before have I waded out to visit it. That can only be seen from my present vantage point as a sin of itself. It's a wonderful big rock, about six feet broad, four feet high, smooth and sloping in undulating layers like fields yielding to the sea, fine and caressing to lie and sun upon.

The small down pillow beneath my head reminds me how much I owe to hard-working forebears, most especially to my mother-in-law who sewed and stuffed it and to her grandmother who years before that plucked the geese to provide the down. My mother-in-law would never have profligated a Thursday afternoon in this way. There are, after all, things I could be doing. There is fruit on the sour cherry trees that the birds are stealing away in alarming quantities this afternoon. I could be up in my kitchen pitting cherries.

The truth is, I crave a little time to waste. I think we need the perspective of a rock in the river, a solitary island from which we can look upstream to where we've been and downstream to where we're going. For years after my mother-in-law died there were Mason jars full of her sour cherries in my freezer. How I wish that instead of the cherries she had hoarded a few July afternoons.

Guestation

THERE MUST BE A REAL VOID IN A HOME WHERE no beloved houseguest ever comes, where no one carries in his suitcases with their mysterious personal contents, where no one brings a deep breath of lives lived in other places.

I am not an exuberant housekeeper. I clean because I can't stand not to clean, and I create order as a resort against insanity. But the ritual of preparing for a houseguest is different. Every corner of the place is seen through the imagined perspective of the guest's eyes and found disturbingly unacceptable.

The fingers fly, the heart races. Now is the time to remove the arm covers from the furniture, to turn the cushions, clean the dead insects from the chandeliers, wipe off the mirrors and the screen doors, wash the porches and trim the lawn. Now is the time to get out the candles, the good linens, the guest towels. Now is the time to deck the house with flowers and hang out the flag.

And then they come—bringing an Amen! to the frantic preparation. The first conversation will inevitably sound like hypocrisy to children who have endured the furious buildup: "I hope you didn't go to any trouble. . . ."

"No, no trouble at all!"

The little boy thinks about his toys stashed away in a closet upstairs, about the mending ready to fall out of the linen cupboard and the clothesbasket under his bed. "They think it's always like this?"

What a child can't know yet is that all the preparation belonged to the magic of anticipation and was no trouble at all, in the normal sense.

The visit passes too quickly as family and guests share some of the everyday and some of the rare. There are tours of the farm, visits here and there, other people dropping in, special foods on the best china, lots of coffee, and conversation early and late.

Then the suitcases go out the front door, there are handshakes and kisses and tearful goodbyes. The visit is over and life settles down.

When I take the dropping flowers out of the "guest room," remove the embroidered sheets from the bed, and cede it to the rightful possessors (who have long since moved in their clutter), then I sense very deeply what the French mean when they say: Partir, c'est à mourir un peu (to part is to die a little).

Yet this was an interlude in daily life to cherish ever afterward. Nothing will be quite the same again, because someone beloved touched here and left behind a small part of himself.

On Parks and People

ORRIN CAME HOME WITH A DOLLAR TEN AND A deflated plastic eightball, which rather well represented the state we were all in following a day at our celebrated amusement park, Cedar Point. A dollar ten seemed quite good until I learned that he'd had nine to begin with. "Nine dollars!" I shrieked. "Why, when I was a girl . . ."

"Yeah, Mom, we know. Nine dollars would have lasted you nine years."

"Never mind," said Dad. "They earned it and they spent it and now it's gone. When they talk about needing money we'll remind them." And so, I suppose, went the conversation in hundreds of other home-going autos.

There is no place in the world (short of a rock concert) where one feels as middle-aged as at an amusement park. Middle age is when the fun house isn't fun anymore, and neither is the front seat of the roller coaster. When there's no longer a need to demonstrate feminine fear or manly courage, what good's a roller coaster? (Have the Fem Libs thought about roller coasters, I wonder?)

And if you happen to get caught up in the youthful gaiety of the place, there are all those age-and-weight-guessing concessions where you pay fifty cents to be told what is perfectly obvious and you'd rather forget.

Riding the Cedar Point and Lake Erie Railroad with soot and smoke blowing in your face is realistic enough to remind you that the "good old days" weren't all that good. I enjoyed the overhead ride on the cable car, but Paul allowed as how it would have been a lot better with snow below and skis to unload.

The ride on a riverboat through an animated wilderness of frontiersmen, Indians, and wild animals impressed me with what a lot of TV-watching zombies we've become. Almost no one demonstrated any appreciation of the heroic efforts at realism. There was nothing in people's reactions to show that they weren't watching the evening weather report. And like most tour spiels, this one had become a cliché to the blond college man in the wilted white ruffled shirt who delivered it over a microphone.

But there is never a cliché in the spontaneous reaction of children to the delights of an amusement park. *This* is the joy of middle age—the vicarious pleasure one takes in the fun of the young: the eleven-year-old who strikes with the mallet and hits the bell three times; the little girl whose eyes and mouth open simultaneously as she gains momentum on the giant slide; the absorption of a thirteen-year-old studying himself in the distortion mirrors of the fun house; the transparent bravery of teenagers on the roller coaster, and their exuberant laughter as they dash to get in line and go again; a three-year-old who smiles then giggles then breaks into laughter on the merry-go-round; the pride on the face of a child as he lives at last his dream of driving something; the triumph that consumes a boy who proves himself braver than his mother at anything.

The ultimate pleasure of the middle-ager at an amusement park is people-watching. The beaches are lined with an odd lot of creatures—fat ladies multiplying their fat while resting their feet, overdressed old men holding hands with younger second wives, disgruntled fathers who would clearly have preferred to spend the day playing golf; harried mothers of too many little children who look as though one more wail of "Mommy!" will send them into nervous collapse.

The teenagers never sit. They always seem to be hurrying to "where it's at" and never finding it. But little boys run, because they know where it's at. It's here and it's now, and if you waste time walking, you'll miss some of it!

There are very young parents trying to conceal from each other the discovery that their children are too young for the fun being forced upon them. They go home tired and disappointed, convinced that they have brought into the world ungrateful monsters. There are other overindulgent families encircling a wailing tyrant and all asking at once, "What *is* it that you want?" (And the people-watchers nod to one another in perfect agreement as to what the kid needs.)

A few young girls are so boldly clad that the most liberal-minded middle-ager is asking himself, "What mother would let a daughter out looking like that!"

And always there are grandmothers who sit placidly watching, acting as home base for a whole brood, guarding the picnic baskets, holding the stuffed animals, mooring a forgotten balloon.

But in the vast majority are a cheerful bunch of pleasant families, with parents resigned to the fact that it's going to be a trying day that definitely belongs to the kids. They stand in lines without complaining, cheerfully dispense dimes and quarters, wipe sticky hands, carry tired toddlers, and lead little girls to the restrooms. They wait patiently at day's end for tardy children, and they do not panic if one is lost. They have faith in other people. They make it possible for me to say, "It was a long, hot, expensive day, but it was worth it!" although I know perfectly well that somebody is going to awaken in the middle of the night and say, "Mama, I've got a stomachache!"

Hotel

It's four-thirty in the morning and I lie awake in the too-light room analyzing my aversion to hotels. The objection of the moment is that they are too cool. With a twinge of revulsion I swing my feet out onto the public carpet and tug the spread from the floor. Something within me resents the unrelenting air conditioning that makes no judgment on the weather. And noisy! The ceaseless hum of the air conditioner is embellished by the hum of the city.

I think then of Carl Sandburg and his reverence for his Chicago—its industry, its clamor and noise, its humanity. No, the city noise I really do not mind. I have lived with noise, and if this noise is different from the crickets, frogs, coon dogs—the country noises I'm used to—the hotel is not responsible.

I look at my watch again, then glance out the window for the source of light that illumines a sixth floor. It is, I discover, a glow prescribed by an architect to carve his monument in the night. There's a drapery to draw, but I have a disdain for drapes.

More than these surface objections are involved with the hollow feeling a hotel room engenders. The smell is alien—stale smoke, booze, dust, furniture wax—what is it?

There lingers here the ghost of orgies past. There is the suggestion of proverbial salesmen and their easy virtue, of teenagers stuffing towels in suitcases, of after-dinner speakers in love with their words, of newlyweds groping naïvely for the secret of life, of entertainers frantic for life-blood acclaim. Something of every sordid novel stalks this room.

Still there's something else, something gruesome in the impersonality of a room decorated for nobody, for a faceless everybody—jarring prints hung in disappointing sameness in every room of every corridor; grotesque lamps with elusive switches; the always-empty drawers; a Bible making its lonely witness. It is the "faceless everybody" who haunts me here, seeking somehow to separate me from my life of meaning.

My spirit senses the loneliness of these everybodies moving, moving, moving through—uprooted for a day or a week or a lifetime from any community that gives them identity.

And I yearn suddenly to escape to a sleeping bag stretched out on the floor of some friendly house, less comfortable, less chic, less private—less alien. Away from these ghosts who condemn me in my narrow morality, my wide hypocrisy, and weep for my shallow compassion.

You Can't Take the Country Out of the Girl . . .

THERE'S NOTHING SO COMPLEMENTARY TO country life as a long intellectual weekend in the city where you are celebrated by dear friends in high-rise apartments. A country husband reads with resignation the signs that say, "It's time to take the girl out of the country."

It is finally accomplished with myriad arrangements and deep sighs of relief. The brief days pass in fulfilling encounters and rich delights.

There are new people with new accents, new ideas, and

new perspectives; there is appreciation for your point of view. There is leisure and luxury and a feeling you could get used to all this! There are dreams made manifest.

Then with the leave-taking and the homeward thoughts, a sense of fulfillment and well-being floods you; and there's a fresh awareness of who you are and where you belong. This has been "delicious," like a small wedge of plum pudding. And you have had *just* enough. Enough of water-chestnuts and wine in the food, enough of Nietzsche and Spinoza and pretentious conversation, enough of remarkable people, and shops full of unique things, enough of blue eye shadow and frosted hair, of sherry and paté, books and the dilettantish life, enough of philosophy and theory, of elevators and buzzers, of objets d'art and Byzantine icons, enough of concrete and steel and the impersonal world of adults. Enough, enough, but what a wonderful "enough" it was. And oh, how you needed it.

Back to windows to wash and laundry to hang, back to milk and boiled potatoes, to white ruffled curtains and third-grade art, hardwares and supermarkets, arithmetic and gym shoes, back to *Time* and the daily funnies, to doors without keys, grass and trees and a neighbor's wave, to little fingers in the cookie batter, blowing hair and the well-scrubbed look, back to battles over prizes in cereal boxes. You didn't miss it five minutes, but oh, how you belong to this!

Happy Ending

If she had been just another pretty teen-ager, I might have forgotten Anne Finseth, but at seventeen she was a "displaced person," a girl who had fallen in love with a life she was losing and had scant hope of regaining. She was returning to Chicago's North Side following a year spent with relatives in Norway. She shared a seat on an air-liner with my husband and me.

During the long hours of our transatlantic flight she

told us of her initial struggles to adjust to life in Norway, of how she had been stranded alone in a strange town and called the only Finseth in the phone book hoping they would be related and take her in. (They weren't but they did!) She told of the difficulty adjusting to the slower pace of life—the tea breaks, for instance, and to a set of values that had more to do with what you were than who you were. She told of how she nearly died of pneumonia fighting the local clothing customs, and of how she finally resigned herself to wearing long heavy stockings and warm underwear.

She was no longer the giddy girl of narrow vision who had left Chicago the previous year with a what-the-heck attitude. She knew it, and she had no appetite for the teenage whirl to which she returned.

When we said goodbye at Kennedy Airport we took her address, hugged her, and wished her well with a great big lump in the throat and an ache in the heart. But she was no correspondent, and for the next few years we just wondered. . . .

On a Saturday afternoon in April four years later a pretty young woman showed up on our doorstep with a very tall Swedish fiancé in tow. It was Anne. She had gone on to college (which wasn't in her plan before Norway) and was now teaching history in Chicago's inner city. At a Scandinavian get-together she had met Manfred, a Swedish farmer here for a year studying American agriculture.

They passed the weekend rhapsodizing over our farm—it was the first time in her life Anne had set foot on a farm! Manfred's father is a potato farmer, so our potato equipment and methods interested him very much. Before they left they invited us to their June wedding at Minnekirche in Chicago's Norwegian settlement.

And so it was that on the second of June at the Scandinavian Singing Society in Chicago, the American farmers joined in a toast to the Norwegian-American bride of a Swedish farmer given by the Irish-American best man.

It's a long way from Chicago's inner city to the farm in Sweden where Anne will spend the rest of her life with Manfred Carlsson. I wanted to write her a book to tell her what it is to be a farmer's wife, but I couldn't wish her anything more symbolic of good luck than did the best man in his Irish toast: "May the road rise to meet you, the wind be always at your back, the rain fall soft upon your fields. . . ."

Again we took down addresses, hugged, and said good-bye, promising to visit them sometime on that farm in Sweden. Again there was a lump in the throat, but the sentiment in the heart was joy!